CAMBRIDGE LIBRARY COLLECTION

Books of enduring scholarly value

Earth Sciences

In the nineteenth century, geology emerged as a distinct academic discipline. It pointed the way towards the theory of evolution, as scientists including Gideon Mantell, Adam Sedgwick, Charles Lyell and Roderick Murchison began to use the evidence of minerals, rock formations and fossils to demonstrate that the earth was older by millions of years than the conventional, Bible-based wisdom had supposed. They argued convincingly that the climate, flora and fauna of the distant past could be deduced from geological evidence. Volcanic activity, the formation of mountains, and the action of glaciers and rivers, tides and ocean currents also became better understood. This series includes landmark publications by pioneers of the modern earth sciences, who advanced the scientific understanding of our planet and the processes by which it is constantly re-shaped.

A Comparative View of the Huttonian and Neptunian Systems of Geology

John Murray (1778–1820) was a public lecturer and writer on chemistry and geology. After attending the University of Edinburgh he became a popular public lecturer on chemistry and pharmacy. He was also a prolific writer of chemistry textbooks which were widely used in British universities. This popular volume, first published anonymously in 1802, contains Murray's critical response to John Playfair's volume *Illustrations of the Huttonian Theory of the Earth,* also published in 1802 and re-issued in this series. In this volume Murray clearly describes both the competing Huttonian and Neptunian (also known as Wernerian) theories of rock formation. Using much of the same geological evidence as Playfair, Murray also objectively analyses the theories' claims through rock and fossil formations and concludes in support of the Wernerian theory. This valuable volume explores one of the major geological controversies of the period and illustrates the main contemporary criticisms of Hutton's work.

Cambridge University Press has long been a pioneer in the reissuing of out-of-print titles from its own backlist, producing digital reprints of books that are still sought after by scholars and students but could not be reprinted economically using traditional technology. The Cambridge Library Collection extends this activity to a wider range of books which are still of importance to researchers and professionals, either for the source material they contain, or as landmarks in the history of their academic discipline.

Drawing from the world-renowned collections in the Cambridge University Library, and guided by the advice of experts in each subject area, Cambridge University Press is using state-of-the-art scanning machines in its own Printing House to capture the content of each book selected for inclusion. The files are processed to give a consistently clear, crisp image, and the books finished to the high quality standard for which the Press is recognised around the world. The latest print-on-demand technology ensures that the books will remain available indefinitely, and that orders for single or multiple copies can quickly be supplied.

The Cambridge Library Collection will bring back to life books of enduring scholarly value (including out-of-copyright works originally issued by other publishers) across a wide range of disciplines in the humanities and social sciences and in science and technology.

A Comparative View
of the Huttonian
and Neptunian
Systems of Geology

In Answer to the Illustrations of the Huttonian
Theory of the Earth, by Professor Playfair

John Murray

CAMBRIDGE
UNIVERSITY PRESS

CAMBRIDGE UNIVERSITY PRESS

Cambridge, New York, Melbourne, Madrid, Cape Town,
Singapore, São Paolo, Delhi, Tokyo, Mexico City

Published in the United States of America by Cambridge University Press, New York

www.cambridge.org
Information on this title: www.cambridge.org/9781108072328

This edition first published 1802
This digitally printed version 2011

ISBN 978-1-108-07232-8 Paperback

A

COMPARATIVE VIEW

OF THE

HUTTONIAN AND NEPTUNIAN

SYSTEMS OF GEOLOGY:

IN ANSWER TO THE

ILLUSTRATIONS OF THE HUTTONIAN THEORY OF THE
EARTH, BY PROFESSOR PLAYFAIR.

EDINBURGH:

PRINTED FOR ROSS AND BLACKWOOD, SOUTH BRIDGE STREET:
AND T. N. LONGMAN, AND O. REES, LONDON.

1802.

EDINBURGH, PRINTED BY MUNDELL AND SON.

PREFACE.

THE prefent publication owes its origin to the
" Illuftrations of the Huttonian Theory of the
Earth," by Profeffor Playfair. In that work this
theory is fo ably fupported, its principles are
placed in fo advantageous a point of view, the
arguments which appear to favour it are fo
forcibly urged, and objections fo ingenioufly,
and often fuccefsfully obviated, that it has given
to the difcuffion of this fubject, an intereft and
form in a great meafure new. To the Author
of the prefent treatife, the Huttonian doctrines,
whatever may be their ingenuity and novelty,
appear vifionary and inconfiftent with the phenomena of Geology; and to a defence of them
fo able, and fo well calculated to convey a favourable impreffion of the general fyftem, he
has been induced by that intereft which every

one feels in the opinions he believes to be juſt, to endeavour to reply. Although, oppoſing the Huttonian Geologiſts, he has been anxious to avoid that ſpirit of hoſtility which too frequently pervades controverſial writings: he has wiſhed to ſtate the arguments on both ſides without partiality, and to reſt the defence of the theory he ſupports on its intrinſic worth.

It will be admitted, that there are few queſtions more calculated to intereſt the ſpeculative inquirer, or more faſcinating, from the grandeur and novelty of the objects it brings before the mind. Nor can it be ſaid to promiſe nothing but the gratification of a vain curioſity. The maxim is too well eſtabliſhed by the hiſtory of ſcience to require proof or illuſtration, that the conſequences which may reſult from any phyſical diſcovery can never be foreſeen, and that no inveſtigation can be deemed unprofitable, which may add to our knowledge of nature. A perfect Theory of the Earth, were it eſtabliſhed, would undoubtedly admit of the moſt important applications; and a ſucceſſion of

[v]

theoretical difcuffions may not lefs contribute to
its attainment, than the accumulation of facts.
Syftems, fays a geological writer, are in the
fciences what the paffions are in the human
mind : they may be the fource of great errors,
but they are the caufe alfo of great exertions.
Either in defending or oppofing them, it is neceffary to obferve with accuracy, to compare and
generalife ; objects apparently minute, acquire
an intereft and importance ; views are fuggefted which often lead to real acquifitions ; facts
are arranged which would have remained ifo-
lated ; and relations traced which would not
have been obferved.

EDINBURGH, December 8. 1807.

CONTENTS.

ERRATA.

Page 5.	line 24.	For Plutonic read	Vulcanic.
13.	14.	gneizs	gneifs.
171.	25.	time	true.

CONTENTS

INTRODUCTION.

THE object of GEOLOGY is to unfold the ftructure of the globe,—to difcover by what caufes its parts have been arranged,—from what operations have originated the general ftratification of its materials, the inequalities with which its furface is diverfified, and the immenfe number of different fubftances of which it is compofed.

Refearches of this kind have by fome been deemed ufelefs, from the fuppofition that their objects cannot be attained. We know not the hiftory of the revolutions of the globe, but we find every where indications of their vaft magnitude and antiquity. We have the moft unexceptionable proof that its whole furface has been

B

covered with the ocean, and that every part of
it has fuffered change : mountains have been
raifed, plains levelled, iflands feparated from continents, and the waters collected fo as to leave an
elevated land. We find it difficult even to conceive caufes adequate to the production of fuch
effects ; and operations fo immenfe feem too remote, from any means of inveftigation we poffefs,
to admit of being explained.

The reply which may be given to obfervations
of this kind is fatisfactory. In any fubject whatever, where the *media* of proof by which a propofition is to be eftablifhed, are imperfectly
known, and where the propofition itfelf is remote
from the more familiar objects of obfervation
and refearch, to inveftigate it is always confidered as impracticable ; and the attempt is treated
as vifionary, until it be crowned with fuccefs.
He who knows not the facts of geology, and the
evidence they afford, may be difpofed to deride
its pretenfions and diftruft its conclufions : but
a more intimate acquaintance with the fubject,
or even a moderate knowledge of the progrefs
of fcience, will teach us to reject fuch narrow
views. Queftions apparently not more within
the reach of inveftigation have been completely
folved ; and man who has weighed the planets,
and meafured their diftances, may prefume to

trace the operations by which the furface of the globe has been arranged.

On making a nearer approach to what at a diſtant view appears ſo difficult, we find that the path we have to purſue is even traced out ; that our inveſtigations are directed by facts which derive not their ſupport from theory, but are eſtabliſhed by the cleareſt evidence. It is prov- ed that the whole furface of the globe has at one period been in a fluid ſtate, and that from this has originated its preſent arrangement. It is certain that this fluidity muſt have been ef- fected by the operation either of fire or of ſome ſolvent. In our farther reſearches, therefore, we are nearly limited to the inquiry, by which of theſe means this effect has been induced ; and from the facts which apparently are within the reach of diſcovery, we need not deſpair of ſolv- ing the problem. The effects of fuſion, and of ſolution, are in general ſo very diſſimilar, or at leaſt the peculiarities of their action are ſo well marked, that from an attentive examina- tion of the properties of minerals, and of the ſtructure and poſition of the great maſſes of the globe, we may hope to diſcover to which of theſe they owe their origin and arrangement ; and if this diſcovery be eſtabliſhed, the labours of geologiſts will be confined to tracing the operation of this agent in producing the phenomena obferved. Their explanations may at firft be imperfect, and their inductions require to be often corrected; but ftill fuch inveftigations muft contribute to the real progrefs of the fcience, and may at length eftablifh a perfect theory.

The fact which has been ftated as the bafis of all geological inquiries,—that the furface of the globe has been in a fluid ftate, is eftablifhed by very ample evidence. In the greater number of the ftrata of the earth, in the moft elevated, as well as in thofe at the greateft depths, fubftances are found in a cryftallized ftate; and even many of thefe ftrata have marks of cryftallization in their entire ftructure. Cryftallization is the arrangement of the particles of a body in a regular determinate form; and it neceffarily implies a previous ftate of fluidity which would allow thefe particles to arrange themfelves in pofitions neceffary to produce thefe forms. Many of the moft folid ftrata contain in their fubftance remains or impreffions of animals and vegetables; and it is obvious, that to admit of the introduction of fuch fubftances, they muft at one time have been, if not in a perfectly fluid, at leaft in a foft or yielding ftate. Laftly, the general difpofition of the materials of the globe, fo far as has been explored, muft have arifen from fluidity, as this only could have arranged them in beds or ftrata parallel to each other, and preferving that parallelifm to a great extent. Thefe appearances are not partial : they extend to the whole furface of the earth, and indubitably prove its former fluidity.

It is obvious that fluidity can be fuppofed to be produced only in two modes. Either the folid matter muft have been fufed by the action of heat, or it muft have been diffolved in fome fluid. Thefe are accordingly the primary propofitions of the different geological theories that have been advanced. Some have fuppofed the furface of the globe to have been foftened or melted by the operation of fire ; others have afcribed the fluidity it muft once have had, to folution; and as there is no fluid in fufficient abundance to have acted the part of a folvent of a quantity of matter fo large, except water, *aqueous folution* has been adopted as the caufe of the original fluidity of minerals, in oppofition to *igneous fufion.*

Thefe oppofite fyftems have been diftinguifhed by the fanciful appellations of the Plutonis and the Neptunian. It would be a fuperfluous tafk to examine the different modifications of them that have been propofed. Of thofe which afcribe the formation of minerals to fufion, it

will probably be admitted that that advanced by
Dr. Hutton is fuperior to any in the extent and
connection of its principles, and in the adaptation of them to the explanation of the phenomena. It will therefore be fufficient to compare the Huttonian theory with the modern
Neptunian fyftem.

PART I.

Statement of the Huttonian and Neptunian Theories.

THE fyftem which Dr. Hutton has traced of the formation of the ftrata of the globe, is calculated to captivate the imagination by its grandeur, and by the appearances which it exhibits of regularity and defign. It prefents a feries of changes, each following from, or connected with the other, and fo nicely adapted, as to be apparently capable of being carried on for an unlimited time. It is claimed as its peculiar excellence, " That it afcribes to the phe-
" nomena of geology an order fimilar to that
" which exifts in the provinces of nature with
" which we are beft acquainted ; that it pro-
" duces feas and continents, not by accident,
" but by the operation of regular and uniform
" caufes; that it makes the decay of one part
" fubfervient to the reftoration of another ; and

B 4

" gives ftability to the whole, not by perpetuat-
" ing individuals, but by reproducing them in
" fucceffion *."

Dr. Hutton conceives that in this globe there
is a fyftem of decay and renovation, and that the
proceffes by which thefe are effected have an u-
niform relation to each other. The folid matter
of the earth is of fuch a nature that it muft be
wafted by the powers which continually operate
upon it. The hardeft rocks are worn down by
air and water: caufes which, however flowly
they may operate, are conftant in their action,
and which therefore, in indefinite time, muft be
equal to the production of the greateft effect.
From the figure of the furface of the earth, the
decayed materials muft be carried towards the
ocean, and ultimately depofited in its bed.
Though this tranfportation may be impeded by
local caufes, or may in general be extremely
flow, yet from the declivity of the land it muft
neceffarily take place, and may therefore be admitted as an uniformly operating caufe.

It is farther affumed, that at great depths in
the mineral regions an immenfe heat is conftantly
prefent, and that this heat operates in the fufion
and confolidation of the fubftances depofited in
thefe regions. It is to the action of this fubterraneous fire that the formation of all our ftrata

is attributed. They confift of the wrecks of a
former world, which have been more or lefs perfectly fufed by this agent, and by fubfequent
cooling have been confolidated.

The fubterraneous fire to which Dr. Hutton
afcribes thofe effects, he conceives to operate at
the fame time under a particular modifying circumftance, which gives a peculiar character to
his fyftem, and frees it from many objections
which were unanfwerable in former hypothefes
of a fimilar kind. This circumftance is compreffion. The fubterraneous fire being placed
at immenfe depths, the fubftances on which it
operates muft be under a vaft preffure. This
will prevent their volatilization in whole or in
part; and from this circumftance it is poffible
to explain appearances in minerals, and qualities they poffefs, which would otherwife appear
inconfiftent with the fuppofition of their being
formed by fire.

The laft part of the Huttonian doctrine is
that which regards the elevation of the ftrata
which have been thus formed; and this likewife
is fuppofed to be the effect of fubterraneous heat.
Its expanfive power muft occafionally be exerted
with very great force; and it is inferred that
this muft have been the caufe which elevated
thefe ftrata, becaufe no other can be affigned
adequate to the production of this effect.

Such is the general outline of the HUTTONIAN THEORY. It is not an attempt to explain the firſt formation of the globe, nor does it proceed on the ſuppoſition, that its preſent is its original ſtate. It reſts on the principle, that, according to eſtabliſhed laws, or from neceſſary changes, operations are conſtantly carrying on, by which the portion of ſolid matter elevated at the ſur-face is worn down, and by which the decay is rendered ſubſervient to the formation of new ſtrata, which will be elevated in their turn. Our world is formed from the decay of one which preceded it, and is now furniſhing the materials from which another is to be raiſed; and this great natural operation is without limi-tation as to time. On the one hand, we find in the ſyſtem itſelf no veſtige of a beginning; on the other, no proſpect of an end.

From this theory are explained the appearan-ces which our ſtrata preſent, in their ſtructure and their poſition. With reſpect to ſtructure, we find conſiderable variety; ſome, as granite, are com-poſed chiefly of ſubſtances in a cryſtallized ſtate, and theſe are ſuppoſed to have been completely fuſed; others, as ſandſtone or chalk, are hete-rogeneous in their texture, or imperfectly con-ſolidated; and theſe are ſuppoſed to have been only in a ſoftened ſtate; and between theſe ex-iſt many intermediate degrees. With reſpect to

poſition, ſome are horizontal, ſome inclined, and
others vertical, irregular, or abrupt; and ſuch
appearances muſt have ariſen from the operation
of that power by which they were raiſed. In
their firſt formation at the bottom of the ocean,
their arrangement muſt have been horizontal;
but in their elevation, by an expanſive power
acting from beneath, their continuity muſt have
been broken, and every variety of poſition produced.

Leſs magnificent in its pretenſions than the
ſyſtem, the outlines of which we have now traced, is the Neptunian Theory, or that which
ſuppoſes the ſtrata of the earth to have had an
aqueous origin. From the appearances which
foſſils preſent, it is inferred that they cannot
have been formed by fuſion; and as ſolution is
the only other mode in which we can conceive
ſoftneſs or fluidity to have been effected, it is
ſuppoſed to have been the agent by which the
matter at the ſurface of the globe was conſolidated. With this ſimple concluſion, it might
be preferable, in the preſent ſtate of geological
knowledge, to reſt ſatisfied, ſince it is proceeding perhaps as far as obſervation can guide us.
To go farther is to wander in the regions of hypotheſis and fancy; and though the opinion thus
deduced might not have the impoſing exterior

of a complete fyftem, it would be more likely to poffefs the ftability of truth.

Such ftrict and cautious induction, however, is not calculated to fatisfy the inquiries of the theorift. It is perhaps neceffary to complete the fyftem ; not merely to reft fatisfied with the proof that minerals have been formed by folution, but to attempt to fhow how this folution, and the fubfequent confolidation from it, have been effected. At leaft, in the prefent difcuffion, this is doing juftice to the Huttonian Theory, as placing both on equal ground.

It may be obferved, however, in proceeding to the ftatement of that modification of the Neptunian fyftem which is generally received, that its author, Werner, has not indulged in hypothefis, but has approached as nearly to an induction of facts as the fubject admits. From the pofitions and connections of the ftrata, he finds reafon to conclude that they have been formed by water. From their relative fituations it is evident that fome have been formed prior to others, and this priority of formation he endeavours to trace and generalife. And, laftly, from their ftructure it is apparent that fome have been chemical, others mechanical depofits, diftinctions which of courfe ought to enter into the fyftem.

The firft principle, then, of the Wernerian theory is, that the materials of which our ftrata

confift were at one period diffolved or fufpended in water, and that from this fluid they had
fuccelïively confolidated in various combinations, partly by cryftallization, and partly by
mechanical depofition. Granite being the rock
which compofes the moft elevated part of the
globe, and which likewife forms the bafis on
which the greater number of the ftrata reft, is
fuppofed to have been firft formed, the different
parts of which it confifts, felfpar, quartz, and
mica, having concreted by a cryftallization nearly fimultaneous. This was accompanied and
followed by the fimilar confolidation of the o-
ther primitive ftrata, gneizs, micaceous fhiftus,
argillaceous fhiftus, porphyry, quartz, &c.

Thefe rocks compofe the chief elevations of
the globe. They are never found to contain
any organic remains, and of courfe their formation muft have been prior to the exiftence of the
vegetable and animal kingdoms.

From the period of the formation of thofe
ftrata, it is fuppofed that the water covering the
furface began to diminifh in height, by retiring
gradually into cavities in the internal parts of
the earth. During this period, other precipitations, chiefly chemical, but in a few cafes partly mechanical, continued to take place. Thefe
formed what Werner terms the intermediate
ftrata, or ftrata of tranfition, of which fome va-

rieties of limeſtone, ſhiſtus, and trap, are the principal. They are incumbent on the primary, and ſometimes, though rarely, contain petrifactions; a proof, however, that organization, or at leaſt the exiſtence of marine animals, had commenced with their formation.

The diminution of the water ſtill continued to proceed, and by the mechanical action of its maſs on the ſtrata formed, it occaſioned in them a partial diſintegration. The materials from this ſource, together with the remaining part of the matter originally diſſolved, by their precipitation and conſolidation formed what are named the ſecondary ſtrata, or the ſtratified rocks, ſandſtone, limeſtone, gypſum, pudding-ſtone, ſome varieties of trap, and various others. Theſe are of a height much inferior to the former. From the predominance of mechanical depoſition, they are arranged generally in horizontal beds, and they are abundant in organic remains; a proof of their formation being poſterior to the full developement of the animal and vegetable kingdoms.

Of theſe different formations the whole ſurface of the globe is formed, at leaſt with the trivial additions of the products of volcanic fire, and the alluvial beds of ſand, clay, and ſoil, ariſing from the waſte of the ſtrata by the waters which run over them.

During the confolidation of thefe ftrata, rents happened in them, by which cavities of very various dimenfions were formed. Into thefe the water, holding various matters in folution, gained accefs, and hence the formation of mineral veins.

Such are the fyftems which we have to compare. In the examination of them, we may confider them under the following divifions: Firft, we may inveftigate the probability of their principles, and the objections to which thefe, *a priori*, are liable: And, fecondly, we may inquire how far they derive fupport from the ftructure and arrangement of the furface of the globe, and from the appearances which minerals actually exhibit.

PART II.

Of the probability of the First Principles of the
Huttonian and Neptunian Theories.

THE first of the series of suppofitions which
compofe the Huttonian theory is, that the ftrata
which form the furface of the globe, are, from
the nature of their conftitution, and the conftant
operation of the agents to which they are expofed, liable to decay; that by being worn down
and tranfported to the fea, they furnifh materials for new ftrata to be formed and afterwards
elevated; and that from the decay of a former
world our ftrata have originated.

To part of thefe fuppofitions fewer valid objections perhaps can be made than to any of the
other principles of the Huttonian fyftem. It
feems to be a neceffary effect arifing from the
nature of the folid materials at the furface of
the globe, that they muft be wafted by the
powers which act upon them; and many facts

have been stated, which establish such a disintegration. The immense beds of sand and gravel
which are found even in elevated situations, afford in particular a clear proof of this kind, since
such substances must evidently have been formed from the waste of solid rocks, and the attrition
of their fragments.

Facts have, however, likewise been stated in
opposition to this principle. Rocks of basaltes,
and of many kinds of granite, it is observed, suffer scarcely any decay, as is evident from the
sharpness of their angles. And from the testi-
mony of history, we have it in our power to as-
certain, that in many places rocks of this kind
occupy the same situation which they did two
thousand years back. These, and similar facts,
however, are to be regarded as exceptions to the
general law; or rather, they serve to prove, that
in such rocks the disintegration is extremely
slow. It seems scarcely possible to deny but
that it must take place to a certain extent; and
as the operations supposed in the Huttonian system are unlimited as to time, it would be in vain
to contest the principle assumed.

It has also been questioned, whether the materials worn down, and conveyed by the rivers,
are carried into the depths of the sea. It has
been alleged, that the greater part of the matter
thus conveyed is thrown by the returning tide

C

upon the fhore and adds to the extent of the land. In many cafes, from local fituation, this is undoubtedly true; but Profeffor Playfair has perhaps fuccefsfully fhown that thefe are exceptions, and that from the declivity of the fhore, the folid matter brought by the rivers muft in general be carried forward, and fpread over the bottom of the ocean. Were this not the cafe, indeed, the increafe of the land, from the accumulation of materials on the fhore, might be expected to be more rapid than it actually is.

Yet though this much be admitted, it is far from eftablifhing the conclufions which Dr. Hutton has deduced, that this difintegration is part of a feries of changes going on in conftant fucceffion, or that it makes part of a fyftem in which a habitable world is always preparing from one exifting, the place of which it is to fupply. If the difintegration be fo flow as is admitted; if, as Dr. Hutton himfelf obferves, the defcription which Polybius has given of the Pontus Euxinus, with the two oppofite Bofphori, the Mæotis, the Propontis, and the port of Byzantium, are as applicable to the prefent ftate of things as they were at the writing of that hiftory; if the ifthmus of Corinth is apparently the fame at prefent as it had been two or three thoufand years ago; if Scylla and Charybdis remain now as they had been in ancient times, rocks hazardous for coafting veffels; if the port of Syracufe, with the ifland which forms the greater and leffer, and the fountain of Arethufa, the water of which the ancients divided from the fea with a wall, do not feem to be altered; and if on the coaft of Egypt we find the rock on which was formerly built the famous tower of Pharos; and at the eaftern extremity of the port Eunofte, the fea-bath, cut in the folid rock on the fhore, to all appearance the fame at this day as they were in ancient times;—if fuch be the extreme flownefs of the difintegration, the reflection is obvious, that, admitting it, a duration will be allowed to the world infinitely beyond our conception, and adequate to any purpofe which we can conceive it defigned to ferve; and there is at leaft no neceffity pointed out for fuppofing an arrangement by which it is to be perpetuated or reftored.

Neither are the facts conclufive which are ftated by Dr. Hutton and Mr. Playfair, to prove that all our ftrata have originated from the wafte of a former world, for they are equally well accounted for by the Wernerian fyftem. It is ftated, that many rocks are found which contain fragments of others, or which are connected with collections of gravel, loofe or confolidated. Such fragments and gravel neceffarily fuppofe the exiftence of former ftrata, from the wafte of which they had originated. It is alfo obferved, that in many of the moft extenfive ftrata of the earth, remains or impreffions of organic fub-ftances are found, both animal and vegetable, and of courfe thefe muft have exifted prior to the formation of fuch ftrata.

Thefe facts are confidered by the Huttonian geologift as fufficient proof of the exiftence of a habitable world, from the decay of which ours has been formed. They are however equally well accounted for by the Neptunift, without admitting fuch a fuppofition. It is fuppofed that the exiftence of marine animals commenced af-ter the cryftallization of the great primary ftra-ta; and that after that period too the waters of the ocean began to diminifh in height, fo as to leave elevated land, on which vegetation com-menced. The retreat of the ocean continued to be gradual for many ages, and during this time the fecondary ftrata were formed. It is obvious, therefore, that the fragments of rock, the fand and gravel which thefe often contain, or with which they are affociated, or which even in many cafes compofe the greateft part of their mafs, might originate from the difintegra-tion of the primary ftrata above the level of the fea; a difintegration to which, in this early pe-riod of their confolidation, they might even be more liable than they now are. And the origin of the remains of marine animals, and even of vegetables, found in the fecondary rocks, it is obvious are equally well accounted for on this theory, fince the exiftence of thefe may have begun previous to the formation of thefe ftrata. The facts, therefore, do not prove the hypothefis of Dr. Hutton, fince on a different hypothefis they are explained with equal facility.

It has been affirmed, however, that the fame appearances of fand and gravel, and of marine impreffions, are occafionally to be met with in the primitive ftrata, and that of courfe the Wernerian explanation is defective; for marine animals are not fuppofed to have exifted at their formation; and it is obvious that the prefence of fand and gravel are true indications of ftrata having exifted before them.

But it is afferted, on the other hand, by Neptunian geologifts, that fuch appearances are not to be met with in ftrata truly primitive, but that when they do occur in ftrata not of the fecondary clafs, it is in thofe of the intermediate kind, or what Werner terms the rocks of tranfition. Thefe, it will be recollected, are fuppofed to be pofterior in their formation to the primary, but prior to the fecondary ftrata, and to have been formed at that period when the exiftence of marine animals, or at leaft of fome fpecies of them, had commenced; and of courfe they may occafionally

be found with impreſſions or remains of theſe
beings. This ſuppoſition is liable to no difficulties, and ſeems to follow juſtly from the facts.
Since certain rocks, having peculiar characters,
and compoſing the moſt elevated parts of the
globe, are found deſtitute of organic remains,
while in others they are abundant, does not this
afford a preſumption, that the former had been
produced prior to the period when theſe beings
began to exiſt? And if rocks are found intermediate in their characters between theſe, connected principally with the primary, but in general leſs elevated, and ſometimes, though rarely, containing veſtiges of ſea animals, is it not
reaſonable to believe that theſe have been intermediate in their formation, and that at leaſt the
few ſpecies of thoſe animals, whoſe remains are
found in them, had begun to exiſt at the time
they were formed?

It will be found that ſuch a ſuppoſition accords much better with theſe phenomena, and
affords a more ſatisfactory ſolution of them than
the hypotheſis by which they are explained in
the Huttonian ſyſtem. According to that ſyſtem, all the ſtrata, both thoſe termed primitive
as well as thoſe named ſecondary, have been
formed from materials depoſited at the bottom
of the ocean from he wrecks of a former world.
They ought therefore all equally to contain or-

ganic remains and impreffions ; and it remains
to be accounted for why thefe fhould be entirely
wanting in many of the ftrata, as in gneifs, or
micaceous fhiftus, while they are prefent in
others. There is only one fuppofition by which
this can be attempted. It is, that the former
ftrata have been in more complete fufion than
the latter, and that thus by the more intenfe
heat, thefe remains, or the impreffions of them,
have been deftroyed. But the explanation is
contradicted by the appearances of thefe ftrata,
the marks of fufion being frequently as complete in thefe which contain fuch remains, as in
thofe which do not. Thus there are many limeftones and marbles containing fhells, in which
the fparry ftructure is as perfect as it is in the primary limeftone, and in which are even diftributed
veins of cryftallized carbonate of lime ; and in
the tranfitive limeftone, the parts which contain
no marine impreffions have no marks of more
perfect fufion than thefe parts in which they are
prefent, nor indeed is there any difference between them. In like manner, the primitive
fhiftus, which has no marks of foreign bodies,
has no appearances of more complete fufion
than the fecondary fhiftus, in which vegetable
impreffions are abundant. The fuppofition,
therefore, of the Huttonian geologift to account
for the abfence of organic remains from thefe

ſtrata, is not merely hypothetical, but inconſiſtent
with the phenomena, while that of the Neptuniſt
is ſo probable, that it ſeems to follow as a corollary from the facts.

The Huttonian explanation does not derive
more ſupport from the other general fact that
has been ſtated in proof of the production of all
our ſtrata from the wrecks of a former world,
—the aſſociation of breccias, of ſand or gravel
with the primary rocks. Several inſtances of
this kind, ſtated by Profeſſor Playfair are extremely doubtful. But granting that there are
ſuch appearances which cannot be diſputed,
they are eaſily explained. It is obvious that on
the gradual retreat of the ſea, which happened
after the formation of the primary rocks, the
moſt elevated would firſt be left bare; and when-
ever this happened, the maſs of water would begin to act upon this dry land, and occaſion in
it diſintegration. It is even reaſonable to believe, though it is not neceſſary to make the ſuppoſition, that this diſintegration would proceed
more rapidly immediately after their conſolidation than when they had been more fully har-
dened by time. Whether this were the caſe or
not, fragments of the elevated rocks muſt have
been worn down into ſand, gravel, &c. and by
the direction of currents, banks of theſe might
be depoſited among the exiſting or forming

ftrata, and might either, according to circumftances, be left in a loofe ftate, or be confolidated by the depofition of the matter ftill diffolved or fufpended in the fluid. There is no
difficulty therefore attending the fact, that fuch
collections are found in the tranfitive, or even in
the truly primitive ftrata, nor does it eftablifh
the conclufion that fuch ftrata had originated
from the wafte of older continents.

There is even a particular phenomenon in
this part of geology, very general in its occur-
rence, which admits of a natural and fatisfactory explanation from the Neptunian theory,
while it requires the moft extravagant fuppofition when confidered according to the principles
of the Huttonian fyftem. The appearance is
that of a bed of breccia incumbent on the primary ftrata, and covered by one or other of the
fecondary ftrata. Over the vertical ftrata, for
example, of the primary fhiftus, there is frequently a bed of breccia, compofed of fragments
of rocks confolidated by fome cement; and over
this bed is a ftratum of fandftone or limeftone.
According to the Neptunian theory, the explanation of this is very eafy. Of thefe the fhiftus,
from its pofition, it is evident muft have been
firft formed : other primary rocks, however, had
been formed at a ftill earlier period, and in pofitions more elevated. As the fea retreated

thefe would be left bare, and by their difintegration would afford the fragments which, depofited on the lower fhiftus, formed the breccia;
and this again, before the retreat of the fea was
completed, fo as to leave it expofed, might be
covered by a depofit of fandftone or limeftone.

The explanation, according to the Huttonian
hypothefis, involves a fuppofition fo extraordinary as to furnifh a fingular contraft with that
of the Neptunian. It is fuppofed * that the
fhiftus had been formed in beds nearly horizontal, and that by an expanfive power exerted
from beneath, thefe had been elevated to the
furface, and placed in a vertical pofition. In
this fituation, the bed of gravel from which the
breccia is formed, had been depofited on the
fummit of the vertical fhiftus. To admit of the
formation of the horizonal ftrata of fandftone,
it is further fuppofed, that the fhiftus, with this
fuperincumbent breccia, had again funk in the
ocean, and remained depreffed for ages, till the
materials of the fandftone were depofited on it.
Thefe materials are fuppofed to have been then
confolidated by the central fire operating on
them, even with the intervention of the deep ftrata of fhiftus on which they are incumbent; and
laftly, we are told that the whole, when thus prepared, were again elevated by a new exertion of

* Illuftrations, &c. p. 52

heat. It may furely be affirmed, without farther reafoning, that fuppofitions fo extravagant and improbable can never be real interpretations of the operations of nature.

We may on the whole conclude, that though the exiftence of organic impreffions and remains may prove that the ftrata containing them have been formed pofterior to the exiftence of the vegetable and animal kingdoms ; and though the prefence of breccia, fand and gravel in the ftrata may likewife prove that other ftrata had exifted, from the wafte of which they had been produced, yet neither of thefe facts eftablifhes the conclufion, that our ftrata have originated from the decay of a former world ; and the explanations of the Huttonian fyftem on this fubject are even lefs probable and fatisfactory than thofe given by the Neptunian theory. The general propofition may therefore be admitted, that the ftrata of the earth are liable to wafte, and that the materials are carried forward to the fea, without proving from any appearances that this makes part of a feries of changes, or is a ftep in the fucceffion of worlds, or that on this planet a habitable world has exifted prior to the prefent, and from the wafte of which this has originated.

The fecond principle affumed in the Huttonian theory, is, that the materials which are collected at the bottom of the ocean are at great
depths expofed to the action of an intenfe heat,
under a ftrong preffure, by which they are fufed
and confolidated, fo as to be capable of forming new ftrata. This principle may be confidered under different afpects; and the difcuffion
of it is important, fince it is capable of affording
a *direct demonftration of the falfity of the Huttonian
hypothefis.*

The difficulties which attend the opinion may
firft be ftated. How is this immenfe heat produced? If from combuftion, the only probable
fource, whence is the combuftible matter derived by which it is excited? Whence is the oxygen fupplied, by which that combuftion muft be
kept up? And how can it be applied to materials collected and depofited from water, while
the operation of that water is at the fame time
excluded? Thefe are difficulties which may be
urged againft this hypothefis, and of which it
will not be an eafy tafk to give a fatisfactory
folution.

The intenfity of the heat required to produce
that fufion whence our ftrata could have originated is beyond what it is poffible perhaps for
the imagination clearly to conceive. This is
evident from the immenfe extent of thefe ftrata.

The higheſt mountains of the globe run in extenſive chains, and theſe, from their connection,
muſt neceſſarily have been formed at one time.
It is not leſs evident, from the difficult fuſibility
of the ſubſtances either compoſing their entire
ſtructure, or contained in them. This may be
illuſtrated by one or two examples.

Carbonate of lime, of which all the calcareous cryſtals, marbles, limeſtone and chalk, conſiſt, cannot be fuſed by any heat which we can
command, or at leaſt can be fuſed only in the
moſt minute quantity. Lavoiſier was unable to
melt a particle of it by the intenſe heat excited
by a burning mirror; and Sauſſure, by the
much more powerful heat excited by the flame
of the blow-pipe, urged by a ſtream of oxygen
gas, was able only to fuſe a particle of it, ſo
ſmall that it required the aid of the microſcope
to diſcover it. What then, ſays Mr. Kirwan,
muſt have been the heat neceſſary to melt whole
mountains of this matter? " Judging from all
" we at preſent know of heat, ſuch a high de-
" gree could only be produced by the pureſt
" air acting on an enormous quantity of com-
" buſtible matter. Now Ehrman obſerved, that
" the combuſtion of 280 cubic inches of air
" acting on charcoal was not able to effect the
" fuſion of one grain of Carrara marble; from
" whence it is apparent, that all the air in the

" atmofphere, nor in ten atmofpheres, would
" not melt a fingle mountain of this fubftance
" of any extent, even if there were a fufficient
" quantity of inflammable matter for it to act
" upon *."

It has been attempted to leffen, if not to obviate the force of this objection. Profeffor Playfair has obferved, that " this reafoning is not
" applicable to Dr. Hutton's hypothefis of fub-
" terraneous heat, becaufe it is grounded on ex-
" periments where that very feparation of the
" volatile and fixed parts takes place, which is
" excluded in that hypothefis. When lime-
" ftone or marble is expofed to fuch a heat as
" is here mentioned, or even to heat of a de-
" gree vaftly inferior, the carbonic gas is ex-
" pelled, and the body is reduced to pure lime.
" The Carrara marble may require a heat of
" 6300 of Wedgewood to melt it in the open
" air, where the carbonic acid gas efcapes from
" it; but under fuch a preffure as would retain
" this gas, it cannot be inferred that it might
" not melt with the heat of a glafs-houfe fur-
" nace †."

This argument, that carbonate of lime may
be fo much more fufible than pure lime, and
that therefore the fubterraneous heat by which

* Geological Effays, p. 453.
† Illuftrations, p. 184.

it is fuppofed to have been fufed, may be very
moderate, fince it aɛted under a compreffion by
which the carbonic acid would be retained, is
fuppofed to be confirmed by an analogical faɛt.
Dr. Black had remarked, that pure barytes is
much lefs fufible than carbonate of barytes,
fince, when carbonate of barytes is expofed to
an intenfe heat, it firft fufes or vitrifies ; it then
begins to part with its carbonic acid, and as it
does fo, it returns to the folid ftate. It is fuppofed, by analogy, that the cafe may be the
fame with carbonate of lime, that when ftrong
compreffion is applied, it may be fufed by a
much lower heat than when the carbonic acid
is allowed to efcape. And fuch a compreffion
is fuppofed, in the Huttonian theory, to have
been applied in the mineral regions, where the
immenfe maffes of carbonate of lime have been
formed.

The faɛt, it may be obferved, on which this
argument is founded, is doubtful. In the mode
of obtaining the barytic earth, by decompofing
carbonate of barytes by heat, the barytes is not
obtained pure ; and if the experiment be made,
as it ufually is, in an earthen veffel, as the decompofition proceeds, the barytes aɛts upon the
earth of the veffel, and forms with it a fpecies
of frit. The fubftance obtained in this way is
both much lefs fufible and lefs foluble than the

pure earth obtained by decompofing the nitrate
of barytes, according to the procefs of Vanque-
lin. This earth is indeed fo eafily fufible, that
the fact cannot be admitted, that barytes, when
pure is more infufible than when combined
with carbonic acid. Precifely the reverfe is the
cafe.

But were it true, the analogy thence extended to lime is not juft. Carbonate of barytes, it
is faid, fufes before it parts with its carbonic acid;
but carbonate of lime does not. No analogy,
therefore, exifts between them in the increafed
fufibility arifing from the combination of carbonic acid with thefe earths ; or there are no
juft grounds for fuppofing that lime, like barytes,
is rendered more fufible by combination with
that acid ; and therefore there is no foundation
for this argument of Profeffor Playfair. And
even if there were, ftill the heat requifite to fufe
carbonate of lime muft be intenfe. In order to
decompofe it, and expel the carbonic acid, a
white heat is requifite, and ftill at this temperature the calcareous carbonate is not fufed ; and
its fufion would probably require a much greater
heat than this. When we confider, therefore,
the great extent of the ftrata of calcareous matter, it is evident, that admitting the above analogical argument to be juft, which it is not, ftill
an immenfe heat muft have been neceffary for

their confolidation, according to the Huttonian hypothefis, a heat with which that of a glafs-houfe furnace cannot be compared.

The original objection may alfo be ftated with equal force, with refpect to other foffils in which there is not the fame mode of eluding it. Quartz, for example, according to the experiments of Lavoifier, and other chemifts, is nearly as infuf-ible as lime. It not only remains unmelted in any heat a furnace can raife, but in the heat ex-cited by a burning mirror. Now there are en-tire mountains of quartz; it is found in large veins in many rocks, and in others, as in granite, it exifts cryftallized as a conftituent part. Thefe, therefore, muft have had an intenfe heat applied to them for their fufion; and no modification, from the prefence of preffure, or any other cir-cumftance that could have any effect in promot-ing the fufion of the quartz, can be imagined. The heat of a glafs-houfe furnace, fays Mr. Play-fair, trufting in the impoffibility of afcertaining the precife fufibility of carbonate of lime, may have been fufficient for its fufion. But there is not the fame uncertainty with regard to quartz. It is calculated by Sauffure, that it requires for its fufion a temperature equal to 4043 of Wedge-wood's pyrometer. Glafs, at a medium, requires only 30 of the fame fcale. From the comparifon of thefe, therefore, we may judge of the intenfi-

D

ty of the heat which would be neceffary to fufe the ftrata of the earth and how far the ftandard of comparifon which Profeffor Playfair points out is juft.

The force of this argument, we may eftimate from the care with which the anfwer to the objection, from the infufibility of carbonate of lime, is repeatedly ftated, and fully urged. In enumerating the difficulties attending the Huttonian hypothefis, from the intenfe heat which muft have been required to fufe certain fubftances, the two examples of carbonate of lime and quartz are the moft obvious, and have been generally ftated together. The defender of that hypothefis, difcovers an analogical argument, by which it is fuppofed, that the force of the objection, from the infufibility of one of thefe fubftances, the calcareous carbonate, may be obviated; and this is brought forward, ftrongly ftated, and repeated under various forms. But the other example, the quartz, is kept out of view No notice is taken of its infufibility having been brought forward equally with that of the calcareous matter, as demonftrating the improbability of this principle of the Huttonian doctrine. Is not this a tacit admiffion of the force of the objection? Where there appears a poffibility of anfwering it, even an imperfect analogical argument, founded on a miftake in point of fact, is not difdained. When, in another inftance, therefore, it is not attempted to anfwer it, is not this to admit, that no fatisfactory reply can be given? that, of courfe, the objection, from the extreme infufibility of foffil fubftances, of which carbonate of lime and quartz are given as the moft ftriking examples, remains in full force? and that the anxiety difplayed in obviating it, in one inftance where it feemed poffible to do fo, is a proof that the objection is not, in the eftimation even of the defenders of this fyftem, trivial, but one which they would wifh, if pof-fible, to remove?

So far, therefore, the improbability of this prin-ciple of the Huttonian theory, that foffils have been fufed by a fubterraneous heat, muft appear evident from the difficulties which attend it. Whether we confider the extent of the ftrata thus fuppofed to have been fufed, or the extreme infufibility of the matter of which they are compofed, the heat requifite muft exceed, in in-tenfity, not only any that we know, but even any that the imagination can conceive ; and for the production of fuch a heat, no adequate caufe can be affigned.

But this argument, however forcible, may ftill be placed in a ftronger point of view. It is the peculiar feature, and as has been ftated by its author, the excellence of the Huttonian

theory, that the operations it fuppofes are intimately connected, and are carried on in a fucceffion to which no limits are placed. It does
not account merely for the appearances which
the earth at prefent exhibits, but unfolds a fyftem in which the deftruction and formation of
fucceffive habitable worlds are explained. In
every part of it, therefore, provifion muft be
made, not merely for the production of an effect,
but for the reproduction of that effect an indefinite number of times. Indulge the defender
of this hypothefis with the affumption, that there
exifted in the bowels of the earth a quantity of
combuftible matter fufficient to have produced
a heat capable of having fufed the prefent ftrata, and that a fufficient quantity of air was fupplied to fupport its burning, even this will
avail him little. Thefe ftrata are fuppofed to
have originated from the decay of others which
exifted before them of a fimilar kind, and which
had been formed in the fame manner; and
they are to be fucceeded by new ftrata likewife
fufed and elevated by heat. Nothing can be
more incontrovertible, than that any accumulation of combuftible matter, or any collection of
air which can be imagined, would not, from the
very principles of the fyftem, be fufficient to
maintain fuch operations. It is a propofition
indeed felf-evident, that a limited quantity could

not ferve to fupport operations, not only immenfe in their extent, but unlimited in their
fucceffion.

On the other hand, Dr. Hutton cannot fhow
that there is any procefs carried on at the furface of the earth fufficiently extenfive to fupply combuftible matter for thefe operations.
He feems to imagine that vegetation is capable
of ferving this purpofe. " Let us fuppofe," fays
he, " the fubterraneous fire fupplied with its
" combuftible materials from this fource, the
" vegetable bodies growing upon the furface of
" the land : Here is a fource provided for the
" fupplying of mineral fire ; a fource which is
" inexhauftible or unlimited, unlefs we are to
" circumfcribe it with regard to time and the
" neceffary ingredients * "

It is true that vegetation is the great fource
of combuftible matter at the furface of the
earth; it ferves as a counterpoife to the various
kinds of oxygenation, decompofes their products,
difengaging the oxygen, and accumulating in
the matter of plants carbon and hydrogen.
Thefe two proceffes appear defigned in the
economy of nature as antagonifts to each other,
oxygenation confuming the combuftible matter which vegetation produces, and vegetation
de-oxydating the products of oxygenation. But

* Theory of the Earth, vol. I. p. 243.

D 3

there is no reafon to believe that the one general procefs exceeds the other in the ultimate effect. It might perhaps be urged as an objection
to the Huttonian theory of the formation of
coal, that it derives that fubftance from the excefs of combuftible matter produced at the furface; for it would require proof that there is
any fuch excefs, or at leaft fuch an excefs as to
continue for an unlimited time. But at any
rate it can never be pretended that from this
fource can be derived that immenfe quantity of
combuftible matter neceffary not only to form
the coal of the fucceeding world, but to fupport that intenfe combuftion which is required
to fufe and confolidate its earthy ftrata.

It even admits of demonftration that no operation carrying on at the furface of the globe
can furnifh an uninterrupted fupply of combuftible matter. In the formation of the ftrata, the
Huttonian hypothefis labours under no difficulty
with refpect to providing materials, becaufe the
fame matter which at one period is fufed and
elevated, is again worn down. But there is no
fuch fucceffion of combuftible matter : in combuftion it is neceffarily rendered unfit to fupport the fame procefs until it be de-oxydated ;
and there is no mean pointed out, or even none
which can be imagined, by which this is effected.

Were even this difficulty removed, it recurs in precisely the same force with respect to the supply of oxygen, which is just as necessary as combustible matter to combustion. No accumulation of oxygen can support combustion for an unlimited time; and no source can be imagined whence a successive supply could be derived.

Professor Playfair seems to be sufficiently aware of the force of these objections, and of the impossibility of supposing the subterraneous heat to be excited and preserved by any species of combustion ; and this, as limiting the discussion, is an important concession. According to all the appearances, he observes, from which the existence of a central heat has been inferred, " it " is of a nature so far different from ordinary " fire, that it may require no circulation of air, " and no supply of materials, to support it. It " is not accompanied with inflammation or " combustion, the great pressure preventing " any separation of parts in the substances on " which it acts, and the absence of that elastic " fluid, without which heat seems to have no " power to decompose bodies, even the most " combustible, contributing to the unalterable " nature of all the substances in the mineral " regions *." Again, " In a region where the

* Illustrations, &c. p. 93.

D 4

" action of heat was accompanied with such
" compreffion as is here fuppofed, there could
" be no fire properly fo called, and no combuf-
" tion. This is admitted by Dr. Hutton ; and
" it is therefore a fallacious argument which is
" brought againft his theory, from the impoffi-
" bility of fire being maintained in the bowels
" of the earth. This impoffibility is precifely
" what he fuppofes, and yet Mr. Kirwan's ar-
" guments are directed not againft the exiftence
" of heat in the interior of the earth, but a-
" gainft the exiftence of burning and inflam-
" mation *."

Profeffor Playfair having thus admitted, in
the moft exprefs terms, that the fubterranean
heat, fuppofed to be the caufe of fuch important
effects in the Huttonian hypothefis, cannot a-
rife from any fpecies of combuftion, proceeds
to explain the mode in which he conceives it
may be excited. His obfervations on this fubject muft be quoted at full length. " It is not
" FIRE in the ufual fenfe of the word, but HEAT,
" which is required for that purpofe ; and there
" is nothing *chimerical* in fuppofing, that nature
" has the means of producing heat, even in a
" very great degree, without the affiftance of
" fuel or of vital air. Friction is a fource of
" heat, unlimited, for what we know, in its ex-

* Illuftrations, p. 182.

" tent, and fo perhaps are other operations, both
" chemical and mechanical; nor are either
" combuftible fubftances, or vital air, concerned
" in the heat thus produced. So alfo the heat of
" the fun's rays in the focus of a burning glafs, the
" moft intenfe that is known, is independent of
" the fubftances juft mentioned; and, though
" that heat certainly could not calcine a metal,
" nor even burn a piece of wood, without
" oxygenous gas, it would doubtlefs produce as
" high a temperature in the abfence as in the
" prefence of that gas.

" It is true, that it is not by the folar rays
" that fubterraneous heat is produced; but
" ftill, from this inftance, we fee, that there is
" no incongruity in fuppofing the production
" of heat to be independent of combuftible bo-
" dies, and of vital air. We are indeed, in
" all cafes, ftrangers to the origin of heat:
" philofophers difpute, at this moment, con-
" cerning the fource of that which is produced
" by burning; and much more are they at a
" lofs to determine, what upholds the light and
" heat of the great luminary, which animates
" all nature by its influence. If we would
" form any opinion on this fubject, we fhall do
" well to attend to the fuggeftions of that great
" philofopher, who was hardly lefs diftinguifhed
" from others by his doubts and conjectures,

[42]

" than by his moſt rigorous and profound in-
" veſtigations. ' May not great, denſe, and fix-
" ed bodies, when heated beyond a certain de-
" gree, emit light ſo copiouſly, as, by the emiſ-
" ſion and reaction of its light, and the reflec-
" tions and refractions of its rays within its
" pores, to grow ſtill hotter, till it comes to a
" certain period of heat, ſuch as is that of the
" ſun? And, are not the ſun and fixed ſtars
" great earths, vehemently hot, whoſe heat is
" conſerved by the greatneſs of the bodies, and
" the mutual action and reaction between them,
" and the light which they emit * ?'

" Some recent experiments ſeem to make the
" ſuggeſtions in this query applicable to an o-
" paque body like the earth, as well as to lumi-
" nous bodies, ſuch as the ſun and fixed ſtars.
" The radiation of heat, where there is no light,
" was firſt rendered probable by the experi-
" ments of M. Pictet of Geneva †; and the
" only objections to which the concluſions from
" thoſe experiments ſeemed liable, are removed
" by the late very important diſcoveries of Dr.
" Herſchel ‡. From theſe it appears that heat
" is capable of refraction and reflection, as well

* Newton's Optics, *ubi ſupra*.
† Eſſai ſur le Feu.
‡ Phil. Tranſ. 1800. p. 84.

" as light, fo that it is not abfurd to fuppofe,
" that *the heat of great, denfe, and fixed bodies,*
" *may be conferved by the greatnefs of the bodies,*
" *and the mutual action and reaction between them*
" *and the heat which they emit *.*"

We may admire the ingenuity of thefe obfervations, in which an argument, obvious and
incontrovertible, is fo enveloped and difguifed,
that it requires fome difcuffion to place it in its
clear light. To what purpofe are the various
fources of heat, enumerated in this reafoning?
To prove that it may exift, or be produced independent of burning. This will be readily
granted : But the reafoning can prove nothing
farther ; it can never warrant the conclufion,
that heat may be afcribed to the operation of
an unknown caufe. If it arife not from burning, it muft be concluded that it is produced by
fome of its other known caufes, by friction, the
fun's rays, fome fpecies of chemical action, or
any other ; and if it can be fhown, that the
fuppofed fubterranean heat cannot originate
from any of thefe caufes, the objection to the
hypothefis which admits it is as ftrong and
irrefiftible, as if combuftion were the fole
origin of heat, and it were proved that it could
not be the effect of that procefs. Now, it can
be fatisfactorily fhown, that any of the known

---

* Illuftrations, &c. p. 186, 187, 188.

caufes of heat, are as incapable of producing it in the interior parts of the globe, to that extent which muft be fuppofed in the Huttonian theory, as combuftion, which, even by its defenders, is confeffed to be inadequate to that purpofe. Hence the general conclufion againft the fuppofition of the exiftence of fuch a heat will be rendered firm and incontrovertible.

Befides, combuftion, friction, and the folar rays, are particularly pointed out as fources of heat. The firft muft be, and is indeed relinquifhed, as the fource of the heat fuppofed to exift in the mineral regions ; and the laft is likewife confeffed to be inapplicable. Friction will not furely be fuppofed to be the fource ; for what caufe can be affigned to excite and preferve a degree of friction capable of producing fuch an effect? All thefe caufes feem, indeed, to be given up, and are apparently introduced only to pave the way for another hypothefis, that "the " heat of great, denfe, and fixed bodies, may be " conferved by the greatnefs of the bodies, and " the mutual action and reaction between them " and the heat which they emit ;" and that to this caufe may be owing the heat in the central regions.

This hypothefis prefents merely an indiftinct idea to the mind, and to be refuted requires only to be analyfed.

The diſtinguiſhing and characteriſtic property of caloric, that from which by far the greater part of its effects ariſes, is its elaſticity, or tendency to exiſt every where in a ſtate of equilibrium. In conſequence of this tendency, it cannot poſſibly be accumulated, and preſerved in that accumulated ſtate, in any body whatever. If a maſs of matter be heated to a high temperature, it immediately begins to part with caloric to the bodies around it, and the ſeparation is proved to take place in two modes ; part of it is thrown off in right lines, or by radiation; and part by a ſlower communication, through the medium of the matter immediately in contact with it. According to the nature of that matter, this communication will be more or leſs ſlow, but it will always take place with a certain celerity ; and even if the body be completely inſulated, or placed in a vacuum, it has been proved, by the moſt accurate experiments, that caloric continues to paſs from it, probably chiefly by radiation, but ſtill ſo as to reduce its temperature.

From theſe facts, which are rigorouſly eſtabliſhed, we are enabled to decide upon this ſubject, and to determine how far the hypotheſis, that the heat of great and denſe bodies may be preſerved by the reaction between them and the heat which they emit, is true.

If we conceive a number of large maſſes of matter at certain temperatures, placed at immenſe diſtances from each other, and either iſolated, or connected only by a very ſubtle medium, ſtill a communication of caloric muſt take place, more or leſs ſlowly, from theſe bodies to each other. A quantity will be thrown off from all of them by radiation ; and if any medium exiſts between them, another portion will be given off by communication. In a certain ſenſe, this caloric will be preſerved amongſt them ; it will be returned from one to another ; and this mutual communication will continue till (if no foreign local cauſe diſturb it) a common temperature is eſtabliſhed among them. But if by ſaying that great and denſe bodies can preſerve their temperature by the mutual reaction between them and the heat they emit, it be meant that any of theſe bodies can, for an *unlimited* time, and without any foreign ſource of heat, preſerve itſelf at a temperature ſuperior to that of the others, we may affirm, in oppoſition even to the authority of Newton, that the concluſion is not juſt, unleſs indeed we ſuppoſe that in ſuch caſes caloric is endowed with peculiar modes of action unknown to us. The defenceleſs Huttonian may ſhelter himſelf under the diſtinction of the ancients, of heat into celeſtial and terreſtrial, and he may confer on either what properties will beſt

accord with his hypothefis; but if we are to follow
the more fober rules of philofophifing, and judge
of the operation of a phyfical agent from what
we know of its powers, it is obvious that in a
fyftem of bodies, whatever may be their greatnefs or denfity, heat cannot be accumulated in
one or more of them for an indefinite time. It
is evident, therefore, that confidering the conjecture of Newton even under this point of view,
it cannot be rendered fubfervient to the purpofe to which the learned Profeffor would apply it.

But the total infufficiency of this hypothefis
is ftill more apparent when we regard the diftribution of caloric, not over a fyftem of bodies
placed at diftances from each other, but merely
over the different parts of one mafs,—the point
of view in which it muft be regarded in this difcuffion with regard to fubterraneous heat. It
is fimply to be repeated, that it is an effential
property of caloric to diffufe itfelf over matter
with which it is in contact, till an equilibrium
or uniformity of temperature is produced,—a
tendency which cannot be interrupted or refifted. Let us fuppofe the central parts of the
earth to be heated to any point; in confequence
of this property, part of that heat muft be diffufed towards the furface in every direction;
and the more denfe the earth is, the more rapid

will this diffusion be; for although the celerity
of the propagation of caloric through bodies is
not precisely according to their densities, it is
more nearly in that ratio than in any other.
No heat, therefore, in the centre of the earth
could be kept accumulated there; its inten-
sity must continue uniformly to diminish; and
as the Huttonian system supposes a succession
of worlds formed by this subterraneous heat, of
which part shall always be solid and habitable,
while another part is fluid, and in a state of pre-
paration to be raised, it is evident, that from the
supposition made of great and dense bodies pre-
serving their heat, the local temperature necef-
sary for these operations cannot be explained.
The heat producing it must come at length to
be equally distributed over the whole mass, and
either render it entirely fluid, or be incapable
of fusing any part.

To this view of the subject, therefore, the
preservation of a central heat in a mass of matter, which is the only one that has any relation
to the present question, the hypothesis of Newton does not apply. That philosopher might
conceive that a great body heated, and placed
at an immense distance from every other, might
from its greatness preserve its heat for a very
long period. Or it might be supposed that a
certain portion of caloric might be preserved in

a fyftem of bodies either opaque or luminous, by the mutual radiation and abforption of it among them. But neither of thefe fuppofitions, were they admitted, contribute in the fmalleft degree to folve the problem, how an intenfe heat could be preferved and accumulated in the central parts of a mafs of matter fuch as our earth. The hypothefis, that great bodies conferve their heat by their greatnefs, and the mutual action between them and the heat they emit, or the eftablifhed fact, that caloric emanates from bodies by radiation, are here of no avail: they have not in fact the moft remote relation to the queftion; and we may admire the ingenuity by which they are apparently connected with it. While caloric continues to be regulated by the laws according to which we know it to operate, it muft, if accumulated in one part of a mafs of matter, diffufe itfelf over the whole; and neither the greatnefs of the body, nor any other property with which this power is endowed, can effectually arreft or counteract this diftribution.

It thus appears that no adequate caufe can be affigned for the production and continuance of that immenfe heat in the bowels of the earth, which is affumed as a firft principle in the Huttonian fyftem, and that every hypothefis on this fubject labours under infuperable difficulties.

E

We may now advance a ftep farther, and fhow by direct *demonftration*, independent of the arguments that have hitherto been urged, that *this principle of the theory is falfe.* It is fuppofed that an intenfe heat always exifts in the central parts of the globe, has exifted, and muft continue active, while this world is regulated by its prefent laws. Let this be admitted, waving every objection as to the mode in which this heat is produced or preferved ; let it be granted that there is a fupply of combuftible matter and of oxygen fufficient for that purpofe, or that great and denfe bodies by their greatnefs conferve their heat ; or let any other fuppofition for the production and continuance of this heat be received which the Huttonian choofes to make ; it is an invariable and effential property of heat to diffufe itfelf over fpace till an equilibrium of temperature is eftablifhed ; and where there is any folid matter as the medium of diffufion, its diftribution is more rapid. If an intenfe heat has always exifted at the central parts of the globe, this heat muft diffufe itfelf towards the circumference, and the diffufion of it muft continue till the whole arrive at a common temperature. The arrangement, therefore, contrived in the Huttonian fyftem for the fucceffive renewal of the habitable part of the globe, and repairing the wafte to which it is fubjected,

is inherently defective. It is always becoming
lefs fit to produce its effects, as the heat at the
centre muft always be diminifhing ; and it muft
come at length to be fubverted by the temperature being rendered uniform over the whole.
The peculiar excellence, therefore, to which
the Huttonian fyftem lays claim, that of pointing out the means of renovation proportioned
to the wafte, and which did it poffefs, would
exalt it over every other, does not belong to it.
In many parts of the economy of nature we
perceive a feries of caufes and effects capable
of producing each other in fucceffion for an
unlimited time, and which, while the fyftem
is governed by its prefent laws, may be faid
to be without any mark of a beginning, or any
indication of an end. Water is raifed by evaporation, forms clouds, defcends again on the earth
in rain, and thus a perpetual circulation of it is
produced. Oxygen is confumed by animals,
and carbonic acid formed ; while, by vegetables,
carbonic acid is abforbed, and oxygen evolved ;
and thefe counteract each other, preferve the purity of the air at nearly an uniform ftandard, and
may be conceived to do fo for any length of time.
The deftruction of the elevated parts of the globe,
the confolidation of the materials thus wafted,
and their elevation in new ftrata, is reprefented
in the Huttonian fyftem, as an operation of a fi-

milar kind; and the view which it thus holds
out was indeed regarded by its author as its
peculiar excellence. It may be admitted to
be one conformable to the general economy of
nature, and, in part, it might appear to be juft,
fince the materials which are worn down are
thofe which are again elevated, and again liable
to wafte. But in the agency by which part of this
feries of operations is fuppofed to be performed,
the fyftem fails. No provifion is made for the
continuance of the fubterranean heat, by which
the matter wafted is to be confolidated and
elevated; and, even allowing its exiftence, from
the nature of this power, it is demonftrated, that
it cannot be preferved for an unlimited time,
in that active and concentrated ftate, which is
neceffary to perform thefe operations, but that
it has a tendency to diffufe itfelf, till it become
every where equal; a tendency which no power
can counteract, and which is utterly fubverfive
of what it is defigned to perform. The fyftem
has within itfelf a principle of decay; its operations muft, of neceffity, have an end, and are
incapable of producing that indefinite fucceffion
of worlds, fuppofed by the Huttonian geologift,
and infeparably affociated with the principles
of his theory.

It is perhaps unneceffary to urge this difcuffion farther. Yet there is another point of

view, under which this fubject may be confidered. Heat, it is fufficiently known, is propagated
through denfe bodies with confiderable celerity.
If a central fire, therefore, of the greateft intenfity, exift, the heat muft be propagated through
the fubftance of the earth towards its furface ;
and this propagation ought to be fuch, that,
even in that period of time of which we have
authentic records, its effects ought to have been
apparent. Yet we have no reafon to believe
that there is any change in the medium temperature of the globe. The climate of particular
countries may be altered, from cultivation, or
other local circumftances, but no important
general alteration appears to have taken place ;
if it had, its effects muft have been confpicuous,
by fymptoms too well marked not to indicate
their caufe. Nay, no change of this kind appears to have happened for a much longer period than that which man has afcertained. It
may be affirmed, that the temperature which at
prefent prevails, is that neceffary for vegetation, animal life, and, in general, for all the
operations of nature; nor could a habitable
world like ours, have exifted with a medium
temperature many degrees inferior to that
which now prevails. The heat at the furface,
therefore, muft have always been nearly the
fame; and, though a central heat has been ex-

E 3

ifting, according to the Huttonian theory, for that immenfe fucceffion of time, during which our world, and others preceding it, have exifted, there has been no propagation of it through the fubftance of the earth. If we can rely on any deduction whatever from the knowledge we poffefs, we may reft affured, that a fyftem involving fuch a fuppofition is falfe ; nor would it be eafy even to imagine any procefs of reafoning by which its falfity could be more clearly demonftrated.

If we compare the Huttonian with the Neptunian theory, it will appear evident, that, according to the one, the prefent arrangement of the furface of the earth admits not of any longer duration, than it does according to the other ; for the principle of renovation, fuppofed in the former, muft have its power extinguifhed, before the wafte, which it is defigned to repair, is complete. The tendency to the deftruction of this globe, as an habitable world, arifes from the difintegration which the ftrata fuffer ; but this difintegration is acknowledged to be extremely flow. " We have mountains " in this country," fays Dr. Hutton, " and thofe " not made of more durable materials than " what are common to the earth, which are not " fenfibly diminifhed in their height with a " thoufand years. The proof of this are the Ro-

" man roads made over fome of thofe hills. I
" have feen thefe roads as diftinct as if only
" made a few years, with fuperficial pits befide
" them, from whence had been dug the gravel,
" or materials of which they had been form-
" ed*." If, in fo long a period, the difintegration is fo inconfiderable as not to be percept-
ible, what muft be required to level thefe
mountains with the fea? Millions of years would
not fuffice. If the comparifon be made, of the
degree of celerity of the propagation of caloric,
and the quantity of matter it has to penetrate
in paffing from the centre to the furface of the
globe, and that of the celerity of the difintegration of the exifting ftrata, as eftablifhed by thefe
facts, it will be evident, that, after ftating the
former as low as poffible, it muft arrive at its termination ; or, in other words, the temperature
muft become uniform in a fhorter period than
that which would be neceffary to complete the
difintegration of the elevated parts of the prefent land.

It is fcarcely neceffary to place the abfurdities
following from this principle in any other point
of view ; yet one more may be added, if poffible
more glaring than the others. The materials
from the wafte of the prefent world are fuppofed

* Theory of the Earth, vol. ii. p. 140.

E 4

to be depofited in the bed of the ocean, and to be there confolidated by the central heat. This heat, therefore, has been capable of propagating itfelf through the immenfe mafs of folid matter interpofed between the centre of the earth and the bottom of the fea, and in a degree fo in-tenfe as to confolidate the loofe matter thus depofited. Yet beyond this it has never been able to advance, even in that vaft period of time in which fucceffive worlds have been forming : In other words, this heat has been diffufed through fome thoufand miles, to the moft intenfe degree, but is there arrefted, and cannot extend itfelf a few miles farther ; a cir-cumftance worthy to be regarded as an interpo-fition of a fuperior power in favour of man. It is thus that the verfatile Huttonian fufpends the laws of nature at his pleafure, or finds them al-ways pliant and accommodating to the principles of his hypothefis.

From the whole of this reafoning, it appears, that under whatever afpect we confider this part of the Huttonian theory, it not only labours under infuperable difficulties, but is proved to be falfe. Whether we fuppofe a vaft accumu-lation of heat, or of matter, capable, by combuf-tion, or any other mode of producing it ; or whether we fuppofe a fucceffive fupply of heat or of matter capable of affording it, ftill the fupply cannot be inexhauſtible, and, of courſe, cannot ſupport the operations it is ſuppoſed to perform, for an indefinite time. And, were it even adequate to this, ſuch is the nature of the agent employed, that it neceſſarily muſt prove deſtructive of the ſyſtem it is deſigned to re-pair. There is an accumulation of proof, which prejudice itſelf, we ſhould be tempted to believe, could not reſiſt, and which is more than ſuffi-cient to eſtabliſh the concluſion, that whatever praiſe may be due to the Huttonian ſyſtem as a ſplendid hypotheſis, it has no claim to the more exalted rank of a juſt theory.

In this diſcuſſion on the exiſtence of a central heat, it has not been thought neceſſary to no-tice the argument for it, drawn from the phe-nomena of volcanoes ; it is ſo obvious that theſe ariſe, not from any ſource of this kind, but from fire excited in the volcanic mountain. Yet, as this is an argument on which ſome ſtreſs is laid, it may not be improper to ſhow its fallacy by a few facts.

1ſt, The heat of the matter erupted from vol-canoes, is not ſuch as it muſt have been, were it derived from that fire which the Huttonian geolo-giſt ſuppoſes to exiſt in the centre of the globe. This heat is ſuppoſed to be ſufficient to fuſe gra-nite, and of courſe quartz, which, according to the experiments of Sauſſure, requires a tem-

perature equal to 4043 of Wedgewood's scale.
But many facts prove, that the heat communi-
cated by volcanic fire, to the matter it throws
out, seldom, if ever, equals 120 of the same
scale. This is evident from shorls, and various
other fossils, fusible at 100, or 110, being found
unaltered in the lava thrown out,—a proof that
they had never been fused ; and lava itself is
fused or vitrified at a temperature below 40.
Such a heat, therefore, can bear no comparison
with that supposed by the Huttonian geologist
to exist in the subterranean regions. It is of
course, a proof, that the lava erupted had not
been derived from that source.

2d, The products of volcanoes are totally unlike those which are supposed in the Huttonian
theory to be formed and thrown out from the
central regions. The latter consist of granite,
porphyry, and trap, substances never rejected
by volcanoes. On the other hand, sulphur is an
abundant production of volcanic fires, while it
is never present in the unstratified rocks ; a
proof of itself decisive that the matter ejected
from volcanoes is not derived from the same
source with that which is supposed to give rise to
the products of the central regions. The stony
matter thrown from volcanoes appears indeed,
from many facts, to be merely the rocks and fossils

of the country, either fufed or partially altered
by the volcanic fire.

Laftly, The extinction of volcanoes fufficiently
proves that they arife merely from the burning,
or mutual chemical action, of a quantity of matter locally accumulated, and fpent after a certain period. If they were connected with the
central regions, no fuch extinction fhould take
place.

The evidence from thefe facts is decifive in
proving that the volcanic mountain is the feat
of the operations the volcano performs; and the
phenomena, therefore, to which thefe give rife,
are no proof whatever of a central fire.

The force of fome of the difficulties which
have been ftated in the courfe of this difcuffion,
and the impoffibility of obviating them, feems
to have been perceived by Dr. Hutton himfelf.
In anfwer to the objections of Mr. Kirwan, he
makes the following obfervations, from which,
however he attempts to difguife it, this is fufficiently evident. " I give myfelf little or no
" trouble about that fire, (the fire neceffary to
" fufe minerals) or take no charge with regard
" to the procuring of that power, as I have not
" founded my theory on the *fuppofition* of fub-
" terraneous fire, however that fire properly fol-
" lows as a conclufion from thofe appearances on
" which the theory is founded. My theory is
" founded upon the general appearances of mine-

" ral bodies, and upon this, that mineral bodies
" muſt neceſſarily have been in a ſtate of fuſion.
" I do not pretend to prove demonſtratively
" that they had been even hot ; however, that
" concluſion alſo naturally follows from their
" having been in fuſion. It is ſufficient for
" me to demonſtrate, that theſe bodies muſt
" have been more or leſs in a ſtate of ſoft-
" neſs and fluidity without any ſpecies of ſolu-
" tion. I do not ſay that this fluidity had been
" without heat, but if that had been the caſe,
" it would have anſwered equally well the pur-
" poſe of my theory, ſo far as this went to ex-
" plain the conſolidation of ſtrata or mineral
" bodies, which I ſtill repeat muſt have been by
" ſimple fluidity, and not by any ſpecies of ſo-
" lution, or any other ſolvent than that univer-
" ſal one which permeates all bodies, and which
" makes them fluid.

" Our author, (Mr. Kirwan), has juſtly
" remarked the difficulty of fire burning be-
" low the earth and ſea. It is not my purpoſe
" here to endeavour to remove thoſe difficulties,
" which perhaps only exiſt in thoſe ſuppoſitions
" which are made on this occaſion ; my purpoſe
" is to ſhow that he had no immediate concern
" with that queſtion in diſcuſſing the ſubject of
" the conſolidation which we actually find in
" the ſtrata of the earth, unleſs my theory with

" regard to the igneous origin of ftony fubftances
" had proceeded upon the fuppofition of a fub-
" terraneous fire. It is furely one thing to em-
" ploy fire and heat to melt mineral bodies, in
" fuppofing this to be the caufe of their confo-
" lidation, and another thing to acknowledge
" fire or heat as having been exerted upon mi-
" neral bodies, when it is clearly proved from
" actual appearances, that thefe bodies had been
" in a melted ftate or that of fimple fluidity.
" Thefe are diftinctions which would be thrown
" away upon the vulgar, but to a man of fcience,
" who analyfes arguments, and reafons ftrictly
" from effect to caufe, this is, I believe, the
" proper way of coming at the truth *."

The obfervations of Profeffor Playfair are to
the fame purpofe : They admit in effect the difficulties attending the fuppofition of a central
heat by which minerals have been fufed, and endeavour to leffen thefe difficulties, on the ground
that mineral fubftances are proved from their
appearance to have already undergone fuch an
operation. " We are not entitled, according
" to any rules of philofophical inveftigation, to
" reject a principle to which we are fairly led
" by an induction from facts, merely becaufe
" we cannot give a fatisfactory explanation of it.

* Theory of the Earth, vol. I. p. 237.

" It would be a very unfound view of phyfical
" fcience which would induce one to deny the
" principle of gravitation, though he cannot
" explain it, or even though the admiffion of it
" reduces him to great metaphyfical difficulties.
" If indeed a downright abfurdity, or inconfif-
" tency with known and eftablifhed facts, be
" involved in any principle, it ought not to be
" admitted, however it may feem calculated to
" explain other appearances. If, for inftance,
" Dr. Hutton held that combuftion was carried
" on in a region where there was no vital air,
" we fhould have faid that he admitted an ab-
" furdity, and that a theory founded on fuch
" poftulata was worfe than chimerical. But if
" the only thing imputable to him is, that, being
" led by induction to admit the fufion of mine-
" ral fubftances in the bowels of the earth, he
" has affumed the exiftence of fuch heat as was
" fufficient for this fufion, though he is unable
" to affign the caufe of it, I believe it will be
" found that his fyftem only fhares in an imper-
" fection which is common to all phyfical theories,
" and which the utmoft improvement of fcience
" will never completely remove*."

Thefe obfervations are fo far juft, that the
merits of any geological theory muft reft on its

* Illuftrations, &c. p. 189.

according with the appearances of minerals. If it fully agrees with thefe appearances, the caufe, thus eftablifhed by induction, ought to be admitted, though it may be liable to difficulties as to its mode of production or operation. But in the application of thefe obfervations to the Huttonian hypothefis there feems to be a miftake, and the reafoning under which its defenders fhelter themfelves as a laft refource, is in feveral refpects incorrect.

It will be recollected, in the firft place, that the term *proof* ufed in this reafoning, is not to be underftood in the fenfe which ftrictly belongs to it, but in a more loofe fignification. When Dr. Hutton fays that he has *proved* minerals to have been formed by fufion, he cannot pretend that his proof is of that kind from which corollaries may be ftrictly deduced, as from a mathematical demonftration. The proof can, from the nature of the fubject, amount to no more than a high probability, and in eftablifhing it, therefore, all the circumftances are to be kept in view, and allowed their due weight. If difficulties attend the conclufion, that minerals have been formed by fubterraneous heat, thefe muft enter into the calculation of probabilities from which the conclufion itfelf is to be drawn. If thefe difficulties be of importance, they may even be capable of balancing probabilities from

the appearances of minerals; or if, on the o-
ther hand, the proof from induction be in any
refpect doubtful or incomplete, fuch difficulties
may be fufficient to juftify the rejection of the
principle partly affumed and partly attempted
to be proved. We are not allowed, therefore,
to draw the conclufion without regarding thefe
difficulties *à priori*, and then fay they are of no
avail, becaufe the proof is already eftablifhed.

But, farther, though it were admitted as really proved, that minerals have been formed by
fufion, it is not this fimple propofition merely
which conftitutes the Huttonian hypothefis; it
is a detailed fyftem, in which a number of fucceffive operations are fuppofed. If, in the feries,
one operation can be pointed out, for the uniform production of which provifion is not made,
the fyftem is fubverted, even though other propofitions which it embraces may be fully proved. From the nature of heat, it can be fhown
that this power cannot be accumulated in the
central parts of the globe, to the extent which
is neceffary in the Huttonian doctrine, and that
there is no fource from which it can be regularly fupplied. Though it were, therefore, proved by indifputable induction from the qualities
of minerals, that they have been formed by fufion, this would not prove the truth of the Huttonian hypothefis. The conclufion muft mere-

ly be admitted, that they have been formed by
fire, but not that the fire has been applied in
the mode, and under the circumſtances which
that hypotheſis points out. Were it even proved, from the poſitions of the ſtrata, that they
had been formed and elevated by a ſubterra-
nean heat, this would ſtill not prove the peculiarities of the ſyſtem, that theſe muſt have
originated from a former continent; that the
materials ariſing from its waſte are ſubjected to
a central heat which at all times continues to
operate, and that a ſucceſſive elevation of them
takes place. In a word, were the igneous origin
of foſſils clearly demonſtrated, it would ſerve
only to eſtabliſh a Vulcanic theory, but would
no more prove the truth of the particular ſyſtem of Hutton, than it would that of Leibnitz
or Buffon. It is to no purpoſe, therefore, that
we are told by Dr. Hutton, that he has proved
that minerals have been formed by fuſion.
This, were it granted, does not prove, nor
even tend to eſtabliſh, the ſeveral propoſitions
which conſtitute his ſyſtem. Some of theſe
are, on the contrary, proved to be falſe; and,
therefore, were the proof of the firſt propoſition
admitted, the refutation of the entire ſyſtem, or
the demonſtration of its falſity would ſtill be
complete.

Laſtly, when the Huttonian doctrine is tried

F

by the teſt which Profeſſor Playfair himſelf has
properly pointed out, the deciſion muſt be giv-
en againſt it. "If," ſays he, "a downright
"abſurdity or inconſiſtency with known and
"eſtabliſhed facts be involved in any princi-
"ple, it ought not to be admitted, however
"it may ſeem calculated to explain other ap-
"pearances." It is a downright abſurdity to
ſuppoſe that caloric could be propagated from
the centre of the earth to the bottom of the ſea,
ſo as to conſolidate the looſe materials there
collected, and ſhould never be able to extend itſelf farther. It is an inconſiſtency with known
and eſtabliſhed facts to ſuppoſe that caloric
may exiſt in an active ſtate in part of a maſs
of matter without being propagated through
the whole; and the principle, therefore, grounded on this ſuppoſition, muſt be rejected, however perfect its explanation of phenomena
might be. The ſuppoſition itſelf is juſt as abſurd as that which the Profeſſor has ſtated, as ſufficient, if it had been maintained, to overturn the theory, that combuſtion
may be carried on where there was no vital
air; for we have not more complete evidence
eſtabliſhing the truth as to the one of theſe
points, than we have as to the other. The fact
is as clearly demonſtrated, that caloric diffuſes
itſelf over matter till an equilibrium of tempe-

rature is attained, as that the prefence of oxygen is neceffary for the procefs of combuftion ;
and any principle which contradicts the one
muft be regarded as falfe, as much as that which
contradicts the other. We may, therefore, pronounce in the words of Mr. Playfair, " that a
" theory founded on fuch *poftulata* is worfe
" than chimerical."

The extenfive difcuffion on this firft principle
of the Huttonian theory, may demand fome a-
pology. Its importance affords one that is fufficient. The evidence for a geological theory,
from the appearances of minerals, muft often
be imperfect or ambiguous, and it evidently is
fo, fince different opinions are held on the fubject. But when the firft principles of a fyftem
are proved to involve abfurdities, and contradictions with known facts, the refutation of it is
neceffarily more complete.

The difcuffion is of importance likewife in the
fubfequent inveftigation. No hypothefis was
ever broached which did not explain, and explain even in a fatisfactory manner, fome of the
phenomena it is defigned to embrace ; and this
will no doubt be found to hold true of the Huttonian theory. But when the falfity of its principles is thus demonftrated, thefe explanations
will be regarded in their true light—as fkilful
or fortunate adaptations of the hypothefis to the

phenomena, and not as juft interpretations of nature.

The laft general principle of the Huttonian fyftem is, that the ftrata after having been fufed and confolidated by fubterranean heat, had been elevated by the fame power. To its probability, *à priori*, the fame objection may be made which has been already urged with refpect to the fufion of the ftrata,—the difficulty or impoffibility of obtaining and preferving a degree of heat fufficient for fuch a purpofe. And were this granted, no principle is pointed out in the theory by which the action of this power can be regulated. It is always reprefented as the peculiar excellence of this fyftem, that none of its operations are the refult of accident, but all are adapted to the attainment of a determinate end. Yet in the fuppofition of this elevation of the ftrata by an expanfive power, no caufe is pointed out for its regular exertion in the mode the theory fuppofes, or why it fhould not be occafionally the caufe of havoc and diforder, as of the renovation of a continent ; and why it might not elevate thefe ftrata before they had been fully prepared. For the production of the effect the Huttonian theory afcribes to it, it is neceffary that it fhould be exerted on the ftrata, confolidated and ready for elevation at the bot-

tom of the fea. Yet there is nothing connected
with it, no circumftance pointed out which
fhould preferve it within thefe limits, or caufe
it to act there more than upon the elevated
land. The principle affumed, therefore, is at
once gratuitous and improbable. How far it
accounts for the appearances which the ftrata
exhibit, is afterwards to be confidered.

We are now to confider the Neptunian theory
with regard to the probability of its firft principle,—that the different foffils have been formed, and the ftrata arranged by depofition from
water.

The great objection to this principle is, the infolubility in water of the matter of which thefe
ftrata confift. The fimple earths which are found
in any confiderable quantity in nature, as lime,
argil, magnefia, and filex, are very fparingly foluble in it; and the compounds which they form
by their union, the different earthy foffils, are
many, or even the greater number of them, apparently infoluble. How therefore can it be
fuppofed that water is the agent which has given
them fluidity, or that they have been confolidated from aqueous folution? " To affirm
" that water was ever capable of diffolving thefe
" fubftances, is to afcribe to it powers which it
" confeffedly has not at prefent; and therefore

" it is to introduce an hypothefis, not merely
" gratuitous, but one which, phyfically fpeaking,
" is abfurd and impoffible *."

In ftrict reafoning, the Neptunift may decline
anfwering objections of this kind. The only
principle which he affumes is, that foffils muft
have been formed by confolidation from aqueous
folution, becaufe the appearances they prefent
are incompatible with the fuppofition of their
being formed by fire. If he can eftablifh this,
his opinion is proved, and his reafoning may
therefore be entirely confined to it. He may
reject, as fubjects beyond the reach of inveftigation, the inquiries refpecting the manner in
which this aqueous folution and the confolidation from it were effected ; and if preffed with
objections from thefe topics, the anfwer is fufficient, that, in the operations of nature, effects
may have arifen from caufes, though in the prefent ftate of our knowledge we may be unable
to point out how from fuch caufes they could
have been produced. The Huttonian geologift,
by afpiring to higher ground, occupies a lefs advantageous pofition. He prefents us with a connected fyftem, in which not only a particular
caufe is pointed out for the production of certain effects, but in which alfo is detailed how

* Illuftrations, p. 18.

and where, with what force, and under what modifying circumſtances this cauſe has operated;
and conſequently objections to any of theſe parts
of his doctrine it is incumbent on him to anſwer.
But the Neptuniſt may reſt ſatisfied with the induction from facts, that water has been the chief
agent in the formation of minerals, and his concluſion will be juſt, though he may be unable
to point out how it has operated.

Let us endeavour to diſcover, however, if the
force of this great objection to the Neptunian
theory may not be obviated.

It may be remarked, that the reaſoning from
which it derives its force is of that faulty kind
which a ſound logic proſcribes. We reaſon con-
fidently from what is at preſent to what has
been, and ſuppoſe unreaſonably that foſſils
muſt have been the ſame at their formation
and arrangement as they now are; that they
muſt have been in the ſame ſtate of aggregation,
muſt have exerted the ſame affinities, and had
the ſame relations to each other. Yet nothing
is more certain than that in theſe reſpects they
muſt have been extremely different, and ſuch
differences muſt have produced the moſt important effects.

Thus it is a principle now admitted by every
chemiſt, that a ſubſtance in maſs, from its ſtate
of aggregation, may be inſoluble in a fluid in

which, when in a ftate of extreme mechanical divifion, it may be diffolved. The examples of this kind in the mineral kingdom are numerous. The corundum ftone, though confifting principally of argillaceous earth, is infoluble in any acid till its cohefion be overcome. Flint is not acted on by an alkali, unlefs it has previoufly been reduced to a ftate of mechanical divifion. The jargon refifts every attempt to decompofe it in the humid way, till its aggregation is overcome by the joint action of potafh and a ftrong heat. The native oxyd of tin is from the fame caufe infoluble in any acid. And in general, in mineral analyfis, this infolubility from aggregation, requires to be overcome by various preliminary operations.

It is apparently from this caufe that feveral of the earths and earthy foffils have been confidered as incapable of being diffolved by water, though foluble in that fluid. This is particularly exemplified in filex, the earth which has been deemed the leaft foluble of any, and of courfe with refpect to which the Neptunift finds the greateft difficulty. When triturated with water, no fenfible portion of it is diffolved; yet there are a variety of facts which prove it to be foluble. If, for example, the filex be combined by fufion with an alkali, and if this compound be afterwards decompofed by an acid, the fili-

ceous earth thrown down is in a ftate of extreme mechanical divifion, into which it cannot
be brought by any other means; and in this
ftate it is foluble in water; fo that if a large
quantity be employed to dilute it, the earth is
not precipitated. Siliceous ftalaⅽtites, though
rare, are fometimes however found, as is admitted by Dr. Hutton himfelf; and in nature, filex
has aⅽtually been found diffolved in many mineral waters. It is fufficient to give as an example of this the water of the Geyfer fountain
in Iceland. A hundred cubic inches of this
water were found by Klaproth to hold diffolved
not lefs than nine grains of filex, and by Dr.
Black 10.8 grains. The latter illuftrious chemift fuppofed that the folubility of the earth
might be promoted by a portion of foda likewife prefent in the water. But Klaproth has
juftly obferved, that the quantity of this is comparatively fo fmall, (only three grains according to his analyfis, and 1.5 according to Dr.
Black's, in 100 cubic inches of the water) that
it is altogether inadequate to the produⅽtion of
fuch an effeⅽt; and that, befides this, the foda
is neutralifed by carbonic acid, while it is only
the pure alkali that promotes the folubility of
this earth. Silex has alfo been found diffolved
in water in which no alkali was prefent *. No

* Kirwan's Geological Effays, p. 117.

doubt therefore can remain of the juſtneſs of Klaproth's opinion, that this earth exiſts in the water of this fountain, from its natural ſolubility in that fluid, affiſted probably by the high temperature, ſince we find it depoſited at the ſides of the fountain ; a depoſition which muſt be owing to the temperature of the ſpring being reduced on expoſure to the air.

It is proved, therefore, that ſiliceous earth is ſoluble in water, and that its apparent inſolubility is owing merely to its ſtate of aggregation. It follows, that the ſame cauſe muſt be affigned for the inſolubility of other foffils; and, of courſe, that, though now inſoluble from their ſtate of aggregation, they may ſtill have had an aqueous origin. To thoſe with whom an example is more convincing than reaſoning, this will be rendered evident by a ſimple fact. The ſiliceous depoſition at the Geyſer fountain, is compoſed, according to Klaproth, of 98 of ſilex, 1.5 of argil, and 0.5 of oxyd of iron ; it is frequently as hard as agate, and is inſoluble in water, as much as any earthy foffil. Yet it owes its formation to aqueous ſolution ; a fact ſufficient to convince us of the error of the concluſion, that foffils cannot be formed by water, becauſe, after their conſolidation, we find them not ſoluble in that fluid. It is, indeed, an actual demonſtration, that ſuch a concluſion is falſe.

Farther, the folvent power of water, with refpect to any fubftance, is invariably promoted
by heat. If we conceive, that, at the commencement of the formation of our ftrata, this
fluid held in folution a great quantity of faline,
earthy, and metallic matters, it is certain that
it would be capable of fuftaining a much higher temperature than pure water would; and
this high temperature, it is indubitable, would
augment its folvent power. It is nearly demonftrated, that, by this agent, the power of combination, of which folution is merely a particular cafe, may be increafed to any requifite extent. The refearches of Berthollet on chemical
affinity, and a number of facts recently eftablifhed, have rendered it extremely probable, that
there are no two bodies in nature which have
not attractions to each other; that water, for
example, has not only an attraction to many
fubftances, but to all; and that, in particular
cafes, exifting attractions are prevented from being efficacious, only by the aggregation in the
bodies concerned being fuperior to the chemical attraction exerted between them, or to other
circumftances preventing their union. The
former power, aggregation, is uniformly diminifhed by heat; and it follows nearly as a
corollary, that, by its action applied in a fufficient degree, any two bodies may be made to

combine, or any fluid may be made the folvent
of any folid. Experiment likewife eftablifhes
this conclufion; for, when water is made to
fuftain a high heat, it becomes the folvent of
fubftances, of which, at a low temperature, it
appears to be incapable of diffolving even the
moft minute portion.

Now, it is no improbable fuppofition, that, at
the period when the materials of the furface of
our globe were in folution, the temperature
may have been much fuperior to that which is
at prefent neceffary for the operations of nature. There muft, neceffarily, have been affigned to this planet, a certain quantity of caloric, and this, before that order was eftablifhed
which now prevails, may have been locally accumulated, and may thus have been capable
of producing, at the furface, the greateft effects.
It has alfo been the opinion of feveral geologifts,
(and, as an hypothefis, there is nothing to prevent it from being affumed) that, at this period,
the atmofphere was not formed; of courfe, the
immenfe quantity of latent heat which it now
contains, would be fenfible and active in the
fluid mafs; and, from this caufe alone, its temperature muft have been high. It is needlefs to
repeat, that, from fuch a caufe, the extent of
which it is not eafy to affign, it is impoffible to
calculate the effects; and he who would limit

them, who would fay that fuch a power could not be prefent, or that it was incapable of the operations fuppofed, would reafon in oppofition to principles eftablifhed by indubitable evidence. This affertion would even ftand contradicted by facts; for, the folution of filex in the Bath waters, Carlfbad waters, thofe of the Geyfer fountain, and other hot fprings, prove the reality of fuch a power in heat, and that it is capable of producing fuch effects.

It is now likewife admitted, that it is impoffible to determine the force of a chemical attraction between any two bodies, or the effect which would arife from its exertion, otherwife than by obfervation or experiment. And the propofition, independent of all theory, would at prefent receive the affent of every chemift, that, if a number of fubftances were brought into contact, in a ftate of extreme divifion, by the medium of a fluid, which itfelf exerted attractions to many of them, it would be impoffible to eftimate what attractions would be moft efficacious, what would be the refult of their concurrent exertion, or what might be fuppofed to be the effect arifing from the prefence of any of them.

To fpeak more precifely with regard to the prefent queftion, it is fuppofed in the Neptunian theory, that at the period of the formation of our ſtrata, there were diſſolved in the chaotic fluid, the different ſimple earths ; the ſimple inflam-mables, the metals ; with various ſaline matters not decompoſed, or if theſe are compounds, the elements of which they conſiſt. To theſe, moſt probably, are to be added the elaſtic fluids, which at preſent form our atmoſphere, and all thoſe principles which are now accumulated in the animal and vegetable ſyſtems. By what power were theſe ſubſtances held in ſolution by the water? The Huttonian, in his argument againſt the Neptunian hypotheſis, replies, that it muſt be ſuppoſed to be by the attraction of each of them ſeparately to that fluid. But the reply is abſurd. Each ſubſtance would exert an attraction more or leſs efficacious to every other preſent, or at leaſt, according to the old chemical notion, it would exert attractions to many of them ; and the effect which would re-ſult from theſe complicated attractions, it is im-poſſible to eſtimate. The number of ſimple ſubſtances found in nature, and which muſt of courſe have been all preſent in this fluid, ex-ceeds forty. Suppoſe the half of theſe to be in ſuch ſmall quantities as to be imperceptible in their action, ſtill, from the numerous attractions of the others, any imaginable effect might ariſe.

To illuſtrate this poſition, if we take one of

thefe fubftances, fuppofe the oxygen, which
now conftitutes one fourth of our atmofphere,
and which enters into the compofition of all
the vegetable and animal fubftances at prefent
exifting in fuch abundance, how are we to determine what may have been the effects refulting from the exertion of its attractions, when
deprived of its elafticity? To fpecify one of
them, it is capable of combining with all the
metals ; and it not only promotes their folubility in water, but increafes the force of their attractions to the earths, or the facility of their
combination with them. How, therefore, are
we to follow it through thefe numerous combinations, fo as to determine the ultimate general effect? Or take hydrogen as an example: it is
capable of combining with fulphur, and with carbon; and thefe compounds again become folvents
of other matters. Hydrogen is even capable, in
its nafcent ftate, of diffolving fome of the metals,
and when deprived of its elafticity, or prefented
in a ftate of condenfation, it is poffible it might
be capable of forming combinations with the
whole of them. Or, laftly, let us take as our illuftration the muriat of foda, which exifts in
fuch immenfe quantity in nature : It is evident,
that if we fuppofe its principles, the acid and
the alkali compounds, as their fimple elements
are unknown to us, we introduce the operations
of fubftances, which, fo far as we know, may

be fully adequate of themfelves to effect the folution of every kind of matter in the chaotic
fluid : Or if, as they have not been decompofed, we fuppofe them fimple, we are not to imagine that they would exift in a mere binary
combination, or as muriat of foda ; their attractions would be divided and modified by the o-
ther fubftances prefent, in fuch a manner, that
he would expofe himfelf to the charge of ignorance of chemiftry, who would venture to fore-
tel the refult. When to thefe three natural
fubftances which we have taken as examples,
the affinities of fo many others are added, it is
obvious, that, from fuch complicated attractions, any effect which implies not a phyfical
impoffibility might be produced.

Another clear illuftration of the fame truth is
afforded by a fingular fact which has been ftated, and which on experiment will be found to
be juft. When the alkaline folutions of filex
and argil are mixed together in equal proportions, " a firm, gelatinous, opalefcent mafs, re-
" fults in a very few minutes. This is perfect-
" ly infoluble in water, yet foluble in acids,
" whether concentrated or diluted, nay even in
" diftilled vinegar, and yet confifts of both fi-
" lex and argil : Here, therefore, the properties
" of the filex muft be confiderably altered *."

* Nicolfon's Journal, vol. iv. p. 543.

No one could have imagined, *à priori*, that argil could render filex foluble in acids, in which
by itfelf it is perfectly infoluble ; and from this
fingle fact it is obvious, that the conclufion is
equally probable, that another fubftance, by the
attraction it exerts to filex, might render it foluble in water. It places, indeed, in a clear
light, the influence arifing from attractions,
which we fhould not have fuppofed would be
important ; and fully proves, that the effects
which may refult from the mutual chemical
actions of a number of fubftances, cannot poffibly be eftimated. It is a fact of the very firft
importance, and which cannot be too highly
prized by the Neptunian geologift.

Reafoning of this kind, though in itfelf fufficiently conclufive, is always rendered more forcible when it can be fhown, that effects, analogous to what are fuppofed, actually exift in nature, and arife from caufes of a fimilar kind.
In the prefent cafe we can obtain the advantage of fuch an illuftration. In the formation
of the animal and vegetable fubftances, affinities are exerted of which we have no knowledge, but from the products that are formed,
and combinations take place which we could
not have imagined, and which we cannot imitate. A few fimple fubftances, carbon, hydrogen, oxygen, azote, and phofphorus, only are

concerned; but fo far from forming the few binary, or ternary compounds which art can produce by combining them, they form an innumerable variety of compounds, diftinguifhed by
the poffeffion of the moft oppofite properties.
To what caufe is this to be referred? The phy-
fiologift may content himfelf with afcribing it
to the fuperintending influence of an unknown
vital power. But if we are not to fatisfy our-
felves with a term deftitute of meaning, and
which, fo far from folving the problem, con-
veys no precife idea, we will fearch for fome
more intelligible theory. We will find it in
the circumftances in which thefe fubftances are
placed. We now know that chemical affinity
is not to be regarded as an abfolute power,
which, in all cafes where it can operate, will
do fo with an equal force. It is moft materially influenced in its action by a variety of circumftances; by force of aggregation, diftances
of the particles, temperature, elafticity or condenfation of the agents concerned, quantity of
the mafs, and others perhaps not afcertained.
Wherever thefe are varied, efpecially where feveral fubftances, having mutual attractions, are
prefent, a difference will take place in the affinities exerted, and thus, from a very flight va-
riation of circumftances, the moft varied combinations will refult. Art cannot eafily pro-

cure or regulate fuch variations, and fhe therefore cannot imitate the operations of animated
nature, in which, by the moft complicated arrangement, the moft ample provifion is made
for the attainment of thefe modifying powers.
Hence, as has been faid, effects refult on which
we could not, *à priori*, calculate ; and the inference from this, to thefe operations by which
the materials of our ftrata were diffolved and
confolidated, is direct and incontrovertible. Attractions muft have been exerted, of which we
know neither the number nor abfolute force ;
and thefe muft have been indefinitely modified,
by the circumftances of temperature, quantity,
condenfation, and others which we cannot determine, and which, in the inftance of the organic products now pointed out, produce fuch
important effects. He, therefore, would reafon
ftrangely, who would abftract entirely thefe circumftances, and tell us, that becaufe when
flint is triturated in a mortar with water it is
not diffolved, it is a phyfical impoffibility that
filiceous earth fhould have been in folution in
the chaotic fluid. His reafoning would be
nearly the fame as that by which it fhould be
affirmed, that fugar cannot be a compound of
carbon, hydrogen, and oxygen, becaufe we cannot form it by combining thefe principles, or

are unable, even by any means, to bring them
into a ternary combination.

Laftly, we do not know what really are the
fimple principles of the fubftances exifting in
the mineral kingdom ; and this of itfelf is fufficient to folve the whole difficulty. If we ana-
lyfe an earthy foffil, we find our analyfis cannot
be carried beyond certain fubftances, which we
term fimple earths : or if we examine the compofition of an ore, we refolve it into a metal combined with fome other body. But can we affirm that thefe earths and metals are certainly
fimple bodies, or that they may not be compounds ? He who knows any thing of the principles of chemiftry, or even of its hiftory, would
never confider the former conclufion as eftablifhed, but would rather incline to the latter,
fince our analyfis can never prove a body to be
abfolutely fimple ; and the progrefs of difcovery
has invariably been, to prove fubftances, apparently fo, to be compounds, and compounds
too of a nature very different from what the
ftate of knowledge prior to the difcovery would
have fuggefted. If fome years back an opinion
on any fubject had been maintained which required the fuppofition that water and air were
compounds, it would probably, from the preva-
lent notion of thefe bodies being elements,
have been rejected, yet the event has fhown

that it would have been juft. And on the fame
grounds, it is equally allowable at prefent to
fuppofe, if it be neceffary to do fo to folve a
difficulty attending a principle eftablifhed on
evidence, that the earths and metals may be
compounds ; and indeed in itfelf fuch a fuppofition is entitled to equal regard with that which
confiders them as fimple, and is perhaps more
conformable to truth. As a general rule to
guard againft extravagant fpeculation, or as a
principle which regulates our arrangements, it
may not be improper to rank every fubftance
as fimple which is not decompofed; but it fhould
not be forgotten, that this is a mere fuppofition,
admitted for thefe reafons, that it is only a
principle of convenience, but that in the abftract it is not more certain than the oppofite
conclufion.

We fhould not indeed err much, perhaps, if
we confidered the greater number of bodies
which are at prefent the fubjects of our knowledge as compounds. Chemiftry is but in
its infancy; within a few years only has it difcovered the compofition of a number of fubftances ; and fhall we believe that it has already
attained the end of its refearches, and that the
varieties of matter which analyfis has difcovered
are truly elementary ? Bodies, as they approach
to fimplicity, apparently become more fubtle,

as is evident from contrafting the fimple gafes
with the compounds they form. When we
compare the amazing fubtility and tenuity of
light, and thofe kinds of matter which give rife
to the phenomena of magnetifm, electricity, and
galvanifm, with the groffnefs and fluggifhnefs
of the metals and earths, or even the fimple
gafes, we will perceive that here the ufual gradations of nature are not obferved; that the
chain which fhould connect material bodies is
as it were broken between thefe different claffes;
and that the fuppofition is not improbable that
thefe groffer bodies are compounds of others
more fubtle, which may approach, or graduate
into thofe kinds of matter that have the moft
undoubted claim to the character of fimplicity.
There are likewife a variety of facts both in
chemiftry and mineralogy, fuch as the apparent
tranfmutation of flint into calcareous earth, and
the production of the earths in the vegetable
fyftem, which appear to indicate that thefe are
compounds.

To fuch fuppofitions, however, though in
themfelves fufficiently probable, and not unfupported, we need not have recourfe. It is fufficient that the abfolute fimplicity of thefe
bodies,—the earths and metals, is not proved.
If the poffibility of their being compounds be
admitted, (and its impoffibility the Huttonian

will never be able to demonftrate) the objection
to the Neptunian theory of their formation, that
they are nearly infoluble in water, is of no force.
It is obvious, that if compounds, their principles
might exift in the fluid or even aerial form, that
they might be foluble in water, or by their mutual attraction might contribute reciprocally to
their folution, and that when combined in other
modes they might form thofe lefs foluble compounds which now exift.

When we confider the circumftances now e-
numerated, the influence of aggregation in preventing folution, the power of temperature in
promoting it, the incalculable effects refulting
from the exertion of complicated affinities, and
the poffibility of fubftances being compounds,
which our imperfect knowledge ranks as fimple,
we can have no hefitation in admitting the conclufion which each feparately eftablifhes, that
foffils may have been formed by water, though
apparently infoluble in that fluid. And if an
induction from facts fhall render probable their
aqueous origin, their prefent infolubility will
form no objection of real force.

Dr. Hutton has likewife fuppofed, that, granting the folubility in water of the matter of which
our ftrata confift, their fubfequent confolidation
cannot be accounted for on the Neptunian theory. The porofity of the mafs could never be

entirely banifhed; and had minerals been confolidated in this way, the folvent ought either
to remain in them in a liquid ftate, or if feparated by evaporation or percolation, it muft
have left the pores empty, and the body per-
vious to water.

" Water," fays Dr. Hutton, " being the ge-
" neral medium in which bodies collected at the
" bottom of the fea are always contained, if
" thofe maffes of collected matter are to be con-
" folidated by folution, it muft be by the dif-
" folution of thofe bodies in that water as a
" menftruum, and by the concretion or cryftal-
" lization of this diffolved matter, that the fpaces,
" firft occupied by water in thofe maffes, are af-
" terwards to be filled with a hard and folid fub-
" ftance; but without fome other power, by
" which the water contained in thofe cavities
" and endlefs labyrinths of the ftrata, fhould be
" feparated in proportion as it had performed its
" tafk, it is inconceivable how thofe maffes,
" however changed from the ftate of their firft
" fubfidence, fhould be abfolutely confolidated,
" without any vifible or fluid water in their
" compofition.

" Befides this difficulty of having the water
" feparated from the porous maffes which are to
" be confolidated, there is another with which,
" upon this fuppofition, we have to ftruggle.

" This is, From whence fhould come the mat-
" ter with which the numberlefs cavities in
" thofe maffes are to be filled ?

" The water in the cavities and interftice of
" thofe bodies compofing ftrata, muft be in a
" ftagnating ftate; confequently, it can only
" act upon the furfaces of thofe cavities which
" are to be filled up. But with what are they
" to be filled ? Not with water; they are full
" of that already : Not with the fubftance of
" the bodies which contain that water; this
" would be only to make one cavity in order
" to fill up another. If, therefore, the cavities
" of the ftrata are to be filled with folid matter,
" by means of water, there muft be made to
" pafs through thofe porous maffes, water im-
" pregnated with fome other fubftances in a dif-
" folved ftate; and the aqueous menftruum muft
" be made to feparate from the diffolved fub-
" ftance, and to depofit the fame in thofe cavi-
" ties through which the folution moves *."

Profeffor Playfair adds to thefe obfervations,
" It is evident, that the confolidation pro-
" duced by the action of water, or of any other
" fluid menftruum, in the manner juft referred
" to, muft neceffarily be imperfect, and can ne-
" ver entirely banifh the porofity of the mafs.

* Theory of the Earth, vol. I. p. 44, 45.

" For the bulk of the folvent, and of the mat-
" ter it contained in folution, being greater
" than the bulk of either taken fingly, when
" the latter was depofited, the former would
" have fufficient room left, and would continue
" to occupy a certain fpace in the interior of
" the ftrata. A liquid folvent, therefore, could
" never fhut up the pores of a body, to the en-
" tire exclufion of itfelf; and, had mineral fub-
" ftances been confolidated, as here fuppofed,
" the folvent ought either to remain within
" them in a liquid ftate, or, if evaporated, fhould
" have left the pores empty, and the body per-
" vious to water. Neither of thefe, however,
" is the fact; many ftratified bodies are per-
" fectly impervious to water, and few mineral
" fubftances contain water in a liquid ftate,
" That they fometimes contain it, chemically
" united to them, is no proof of their folidity
" having been brought about by that fluid;
" for fuch chemical union is as confiftent with
" the fuppofition of igneous as of aqueous con-
" folidation, fince the region in which the fire
" was applied, on every hypothefis, muft have
" abounded with humidity *."

Thefe objections, it may be affirmed, arife
from afcribing an hypothefis to the Neptunian

* Illuftrations of the Huttonian Theory, p. 16, 17.

theory which does not belong to it, and that, with refpect to this point, it labours under no real difficulty. This will be evident, from ſtating briefly the modes in which it ſuppoſes conſolidation to take place from fluidity by water.

If the ſolid ſubſtance be completely diſſolved in the water, it may ſeparate from it by cryſtallization, this cryſtallization being determined by a change of circumſtances in the ſolution. If the ſolubility of the ſolid had been augmented by the heat of the ſolvent, or by the preſence in it of certain principles, ſuppoſe of an aerial kind, it is evident, that on a reduction of temperature, or on the eſcape of theſe principles, or even on new combinations taking place, cryſtallization would commence : it would proceed more or leſs rapidly according to the extent in the change of circumſtances bringing it about ; and in this manner might be formed the different ſtrata which exhibit marks of cryſtallization in their ſtructure. No one can doubt that in this way maſſes of the greateſt hardneſs and denſity might be formed, and the chemiſt indeed has daily opportunities of obſerving the hardneſs which a cryſtalline maſs acquires, when it ſtands for ſome time with the liquor over it from which it had been depoſited, every vacuity which may have been left in it at its firſt cryſtallization being filled up by ſubſequent cryftalline depofition, and the fluid filling thefe cavities, being extruded.

Other ftrata may have been formed merely by depofition of folid matter, not diffolved, but fufpended in a fluid. In this manner have apparently originated a confiderable number of the fecondary ftrata. The particles, depofited in large beds, would at firft have little cohefion; but there is no improbability in the fuppofition, that in this foft ftate the attraction of aggregation would be exerted between them, fince we know that neither this fpecies of attraction, nor chemical affinity, is entirely confined in its action to the ultimate particles of matter, but can be exerted between minute maffes. The particles would thus approximate, the interpofed fluid would be forced out, and a mafs more or lefs compact would be formed. This, on remaining a long time at reft, might attain a great degree of hardnefs; and it may be obferved, that in nature no foffil, at leaft none uncryftallized, is fo perfectly denfe as to be altogether impervious to water. This fpecies of confolidation would be ftill more promoted if a chemical were mixed with the mechanical depofite, or matter was feparated that had been diffolved, at the fame time with the fubfidence of the particles fufpended, a circumftance which appears to have often happened.

The futility of Dr. Hutton's objections to this account of confolidation from aqueous folution will now be apparent. They do not indeed apply to it, but to an hypothefis which no Neptunift maintains. In accounting for the confolidation of the ftrata by water, it is not conceived that the loofe materials are firft depofited at the bottom of the fea, and that then fome matter is introduced into their pores by which they are rendered hard. They have been formed either by cryftallization or precipitation. In the former mode it is obvious that no operation of this kind can be fuppofed requifite to confolidate them; in the latter cafe it is in general equally unneceffary, becaufe when the particles in a ftate of minute divifion were depofited, they might cohere, and would not, as Dr. Hutton conceives, require the introduction of any other matter to complete their confolidation. But were even fuch a fuppofition made, the objection would be groundlefs. If the water fuppofed to be introduced into the precipitated mafs had particles of folid matter merely *fufpended* in it, thefe would be difpofed to fubfide, and cohere more or lefs firmly with the folid matter into which the water had been infinuated, and the water itfelf would at the fame time be extruded. Or if the folid matter were *diffolved*, it would be difpofed to cryftallize, by obtaining a nucleus

in the ſtrata into which the water containing it
had infinuated. There are even facts which
fully prove that ſtony particles may be introduced by infiltration into ſtrata, and may give them
additional hardneſs *, a cauſe which, it is not
improbable, has operated to a conſiderable extent in mineral conſolidation.

Still Dr. Hutton repeats his objections to the
Neptunian explanation of conſolidation. In
his reply to Mr. Kirwan, who had ſtated this
explanation, he has the following ſingular obſervations: " If I underſtand our author's (Mr.
" Kirwan's) argument, the particles of ſtone
" are, by their mutual attractions, to leave thoſe
" hard and ſolid bodies which compoſe the
" ſtrata, that is to ſay, thoſe hard bodies are to
" diſſolve themſelves; but, To what purpoſe?
" This muſt be to fill up the interſtices, which
" we muſt ſuppoſe occupied by the water. In
" that caſe, we ſhould find the original inter-
" ſtices filled with the ſubſtances which had
" compoſed the ſtrata, and we ſhould find the
" water tranſlated into the places of thoſe
" bodies; here would be properly a tranſmuta-
" tion, but no conſolidation of the ſtrata, ſuch
" as we are to look for, and ſuch as we actually
" find among thoſe ſtrata. It may be very

* Kirwan's Geological Eſſays, p. 132.

[95]

" eafy for our author to form thofe explanations
" of natural phenomena; it cofts no tedious
" obfervation of facts, which are to be gathered
" with labour, patience, and attention; he has
" but to look into his own fancy, as philofo-
" phers did in former times, when they faw
" the abhorrence of a vacuum and explained
" the pump. It is thus that we are here told
" the confolidation of ftrata *arifes from the mu-*
" *tual attraction of the component particles of*
" *ftones to each other.* The power, by which the
" particles of folid ftony bodies retain their
" places in relation to each other, and refift fe-
" paration from the mafs, may, no doubt, be
" properly enough termed their mutual attrac-
" tions; but we are not here inquiring after
" that power; we are to inveftigate the power
" by which the particles of hard and ftony
" bodies had been feparated, contrary to their
" mutual attractions, in order to form new con-
" cretions, by being again brought within the
" fpheres of action in which their mutual at-
" tractions might take place, and make them
" one folid body. Now, to fay that this is by
" their mutual attraction, is either to mifunder-
" ftand the proper queftion, or to give a moft
" prepofterous anfwer *."

* Theory of the Earth, vol. i. p. 228, 229.

The caufe of this fingularly confufed and miftaken ftatement, appears to be, Dr. Hutton having always before him a limited view of his fubject. When he afks, by what caufe are the ftrata confolidated? he appears to mean, not as his expreffion would imply, the ftrata in general, but thofe particular ftrata which confift of mechanical fragments cemented together. He confiders thefe ftrata as compofed of particular collections of fubftances, limeftone, for example, of remains of marine animals, and pudding-ftone, of pebbles or gravel, which are confolidated by being fixed in a common cement. When, therefore, his queftion is anfwered, by faying, that confolidation is owing to the mutual attraction of the particles, the anfwer appears to him prepofterous. But this is owing to the prepofterous manner in which he has conceived the fubject. It is a proper anfwer with refpect to the confolidation of homogenous ftrata, which amount at leaft to nine in ten of the whole. And, if the queftion were farther put, with refpect to thefe particular ftrata, which are compofed of mechanical fragments cemented together, the anfwer of the Neptunian would then be, that the cementing matter of thefe ftrata had been depofited around thefe fragments from folution or fufpenfion in water, and that the attraction of aggregation being exerted between its particles, it had confolidated, and, of courfe, had confolidated the whole mafs.

Or rather the limited view which Dr. Hutton takes of this fubject, arifes from his hypothefis, that all the ftrata have arifen from the decay of former ones; and that, therefore, their materials have once confifted of mechanical fragments loofely depofited. He fhould have recollected, however, that this fuppofition is not admitted by his antagonift, and that he therefore was not at liberty to attribute it to him, or to reafon from it as if it were received. The opinion of the Neptunift as to the original condition of the materials of the ftrata is altogether different, and it is not to this opinion, but to an hypothefis, the creature of his own imagination, and of his limited views, that the objections of Dr. Hutton on confolidation apply.

Examples of both the kinds of confolidation pointed out, are abundant in nature. Calcareous cryftals and ftalactites, and even filiceous ftalactites, are frequently formed by water at the furface, and fhow, at leaft, that from cryftallization, perfect confolidation may take place; a conclufion, indeed, which no one can doubt. Of depofition, too hafty to admit of cryftallization, the filiceous incruftation at the Geyfer fountain is a fufficient example. Its hardnefs is frequently fuch as to be equal to that of the

H

more hard filiceous foffils, agate, or chalcedony.
Many other examples of confolidation from thefe
caufes, as well as from mechanical depofition,
are ftated by Mr. Kirwan *. One of thefe may
be felected,—that of the regeneration of granite,
or the formation of a hard ftone from the materials of that foffil. It occurred in a mole con-
ftructed in the Oder, the fides of which were of
granite, and the middle fpace was filled with
granitic fand. This concreted into a fubftance
fo hard as to be impenetrable by water. The
mode in which Dr. Hutton attempts to obviate
this fact could fcarcely be explained but in
his own words : " Here is an example, accord-
" ing to our author, of granite formed in the
" moift way. But now I muft afk to fee the
" evidence of that fact ; for, from what our au-
" thor has told us, I do not even fee reafon to
" conclude, that there was the leaft concretion,
" or any ftone formed at all. A body of fand
" will be *fo compacted as to be impenetrable by
" water*, with the introduction of a very little
" mud, and without any degree of concretion.
" Muddy water indeed cannot be made to pafs
" through fuch a body without compacting it
" fo ; and this every body finds to their coft,
" who have attempted to make a filter of that
" kind †."

* Geological Effays, Effay iv.
† Theory of the Earth, vol. i. p. 267.

This requires no anfwer. It is admitted, that the fubftance thus formed, was fo hard as to be impenetrable by water. (Mr. Kirwan has fince added, that it was even fo compact that it could fcarce be feparated from the real granite to which it was contiguous, by a blow). Yet Dr. Hutton ventures to tell us, that this hardnefs is not from the particles having concreted, but from their having been compacted by mud. In another part of his work, he gives diftinctions, which, he informs his readers, are not for the vulgar to underftand. This diftinction between *compacting* and *concreting*, is perhaps one of this kind, and he who underftands it fo as to apply it to the folution of the prefent difficulty, may congratulate himfelf on being none of the vulgar, but probably a profound logician of Dr. Hutton's fchool.

If foffils have been confolidated from water, it might be fuppofed that they fhould contain a portion of that fluid. The anfwer is, that they actually do fo. No aggregate perhaps exifts in nature, which does not lofe weight on expofure to heat, from the efcape of water; fome of the moft folid rocks lofe confiderably, and feveral of the hardeft minerals contain 5 or 10 parts in the 100. The compofition of foffils, therefore, fuffi-ciently accords with the fuppofition of their

being confolidated by concretion from water;
nor has it been fhown that they ought to contain more water than they actually do. In all
cryftallized bodies, the quantity retained is different, and is dependent partly on their attraction to the fluid, and partly on circumftances
attending their cryftallization. And there are
fubftances actually depofited from water which
do not contain any fenfible portion of it :
Such, according to the analyfis of Klaproth, is
the filiceous incruftation of the Geyfer fountain.

We may now contraft thefe two fyftems with
refpect to the probability of their firft principles,
and compare the explanation which either affords, with that which the nature of the fubject
admits. We have feen, not only that the Huttonian fyftem labours under infurmountable
difficulties, in accounting for the production of
that heat by which it fuppofes the ftrata to be
fufed and confolidated, but it has been proved,
that its principles ftand in direct oppofition to
eftablifhed facts, and are therefore falfe. The
Neptunian fyftem, again, in accounting for the
original fluidity of the furface of the globe from
the operation of water, may appear to be liable
to fome difficulties ; but it is fertile in refources,
fince caufes can be pointed out, by which the

action of that fluid might be modified, fo as to produce the effects afcribed to its operation. It may be impoffible to determine which of the caufes pointed out did actually operate, or to which the greateft fhare in the ultimate effect is to be attributed. But what is to be expected from the geologift in fuch refearches? not furely that he is to point out the precife manner in which thefe immenfe operations have been conducted, or, as if he had been a fpectator of them, defcribe them in detail. The theorift may make pretenfions of this kind, but, under the name of a fyftem, he will give us merely a dream. It is fufficient if, from induction, a particular caufe be eftablifhed, and if this caufe be rendered probable, or objections to it obviated by reafoning which involves no improbable fuppofitions. This is attained in the Neptunian theory. From the appearances of minerals, we fhall find fufficient reafon to conclude, that they have been formed by water; and the objections which might, *a priori*, be made to this conclu-fion, from the powers of water being inade-quate to fuch effects, we have feen are fufficient-ly obviated, by the admiffion of certain modi-fying circumftances, which not only might have operated, but which muft have done fo to a certain extent. There is, therefore, nothing to prevent the admiffion of the principle which induction eftablishes.

On this fubject too, were it neceffary, the principle might be infifted on, that, if a certain caufe be fuggefted by the phenomena, it ought to be admitted, though there are difficulties in explaining its precife mode of operation. Dr. Hutton and Profeffor Playfair apply this argument to the fupport of their doctrine, without perceiving or acknowledging that it can be brought with equal force to eftablifh the Neptunian theory : " I have proved," fays Dr. Hutton, " that thofe ftony fubftances have been " in the fluid ftate of fufion ; and from this I " have inferred the former exiftence of an in- " ternal heat, a fubterraneous fire, or a certain " caufe of fufion, by whatever name it fhall be " called, and by whatever means it fhall have " been procured. Though I fhould confefs my " ignorance with regard to the means of pro- " curing fire, the evidence of the melting ope- " ration, or former fluidity of thefe mineral " bodies, would not be thereby in the leaft di- " minifhed *." It is obvious, that the whole of this argument may be adapted with equal juftice to the Neptunian theory. I have proved, (a defender of that fyftem might fay) that thefe

* Theory of the Earth, vol. i. p. 239.

ftony fubftances have been in the fluid ftate of
folution ; and from this I have inferred the exiftence of a folvent, or a certain caufe of folution, by whatever name it fhall be called, and
by whatever means it fhall have been procured ;
and though I fhould confefs my ignorance with
regard to the means of procuring this folvent,
the evidence of the diffolving operation, or former fluidity of thefe mineral bodies, would not
be thereby in the leaft diminifhed.

This mode of reafoning can indeed be employed with much more force and juftice by
the Neptunift than by the Huttonian ; for, as
has already been fhown, though it were demonftrated that minerals had been formed by fufion,
this would not eftablifh the truth of the Huttonian fyftem, becaufe that fyftem involves principles phyfically impoffible, and which therefore no induction can eftablifh ; while on the
contrary, the Neptunian is at moft liable to fome
difficulties which are eafily removed, and which,
though they were not, would not be fufficient
to invalidate the evidence which the appearances
of minerals afford. Little doubt, therefore, it
is prefumed, can remain of the decided fuperiority of the firft principles of the one to thofe of
the other. We have next to examine what fupport thefe theories derive from the phenomena or
geology. The arguments of this kind which

H 4

have been advanced may be confidered under
two claffes; thofe drawn from the pofitions and
connections of the ftrata of the earth, and thefe
from the properties and appearances of individual foffils.

PART III.

Of the Arguments in support of the Huttonian and Neptunian Theories, from the Positions of the Strata of the Globe.

THE immense masses which compose the surface of the earth are extremely various in their positions. Some are altogether irregular, others are disposed in beds or strata which are vertical, horizontal, or inclined at different angles to the horizon, varying in their extent and thickness, alternating often with each other, and frequently intimately connected at their junctions. These are facts of the first importance in geology, and which require to be explained in any theory of the earth.

The Huttonian explanations of the formation of the stratified rocks, and of that of the unstratified, are very different from each other, and require separate consideration.

Of the STRATIFIED ROCKS it has juſt been obſerved, that ſome are vertical in their poſition,
ſome horizontal, or nearly ſo, and others inclined
at different angles to the horizon. Dr. Hutton
ſuppoſes that all theſe have originally been horizontal, that the matter of which they confiſt
has been ſpread out on the bottom of the ocean,
conſolidated by fuſion by ſubterraneous heat,
and afterwards elevated by the expanſive power
of that heat, ſo as to aſſume the poſitions they
are found to have.

In the Neptunian ſyſtem it is ſuppoſed that
the poſitions of the ſtrata have been determined,
partly by the figure of the baſe or ground on
which they have been depoſited, and partly by
their depoſition having been a ſpecies of cryſtallization. The primitive ſtrata, which are
generally vertical or highly inclined, had firſt
been depoſited from the original fluid in which
their ſubſtance was diſſolved, and they had
taken the form in which we find them, by having been haſtily cryſtallized. From theſe immenſe maſſes the water is ſuppoſed to have
retired into the more hollow parts of the globe,
or into caverns in its central parts. This retreat was ſo gradual as to have continued for
ſeveral ages, and during its continuance, the
ſecondary ſtrata appear to have been formed by depoſition of matter from the ocean,

on the fides of the vertical and highly inclined
ftrata already formed : and laftly, by the continued retreat of the fea, thefe fecondary ftrata
have likewife been left expofed.

It is objected to this account of the formation
of the ftrata, that it does not account for their
inclined or vertical pofition. If matter, it is
faid, was depofited from a fluid, it would arrange
itfelf in a horizontal bed; the inclination of the
bottom or ground on which it is depofited
might have fome effect in modifying the pofition, but this would foon ceafe; the matter precipitated, from this inclination, muft be depofited
unequally ; being carried forward where the inclination is greateft, it would tend to arrange
the whole horizontally, and the different beds
or ftrata, as they are formed, would approach
more and more to that pofition. To what caufe
then are to be afcribed the vertical, or the highly
inclined pofitions which they frequently have,
and the parallelifm which they preferve with
great exactnefs under very extenfive and varied
incurvations ?

The force of this objection depends entirely
on the fuppofition, that in the depofition of thefe
ftrata, their matter had been previoufly merely
fufpended mechanically in a fluid, and had fub-
fided by reft. Had this been the cafe they muft
have been arranged in horizontal beds, and ought

of courfe to be found in that pofition. The opinion maintained in the Neptunian theory however is, that they had been chemically diffolved,
and had feparated and concreted by a fpecies
of cryftallization. Thefe cryftalline depofites
would be in large irregular maffes, as granite,
the rock of primary formation, is; and the fluid
ftill continuing to depofite matter by cryftallization, this matter, in conformity to the laws of
that procefs, would cryftallize on the fides of the
maffes already produced; and thus the appearance of vertical ftrata would be formed: or the
divifion of thefe might even be determined by
the procefs of cryftallization itfelf. With regard to the finuofities and incurvations in thefe
ftrata, it may be difficult to point out precifely
how cryftallization fhould produce them. But
we have fufficient evidence from facts to prove
that this procefs is capable of producing fuch effects, and actually does produce them in cafes fo
obvious that there is no room for doubt. Sauffure has obferved, with a view to the illuftration
of this opinion, which he maintains, that thefe
very appearances are found in ftalactites and ala-
bafter. Thefe confift of different layers, which,
inftead of being in a ftraight line, are found
with all thefe varieties of bending and waving
which in the ftrata are confpicuous on a large
fcale. For the production of thefe no juft caufe

can be affigned but the cryftallization, imperfect as it is, by which thefe fubftances are formed, a caufe by which not only the figure, as
Dr. Hutton has alleged, but likewife the ftructure of bodies, is determined. It is obvious,
that fince this is proved to be a fufficient caufe,
the reafoning is nugatory which is founded
merely on the difficulty of conceiving *a priori*
how it fhould produce thofe effects, becaufe the
anfwer is fufficient that it actually does fo. And
although we may be unable to fay how cryftallization fhould give rife to " the inflexion of the
" ftrata, the fimple curvature which they affect,
" and that parallelifm of their layers which in
" all their bendings is fo accurately preferved*,"
yet we fee in the example quoted by Sauffure
that it can produce all thefe appearances, and
therefore it may, on a larger fcale, be adequate
to their production in the mountains of the globe.
The difference of magnitude cannot alter the
nature or powers of the procefs employed.

In reality, fome caufe of this kind muft even
be had recourfe to by the Huttonian geologift.
It is fuppofed indeed that the ftratification has
been the effect of depofition from water; but
this cannot be admitted ; for although the materials of thefe ftrata might at their depofition be

* Illuftrations, p. 232.

conceived in the Huttonian theory to be arranged in horizontal beds, yet their fubfequent
fufion, which in thefe primary rocks is fuppofed
not to have been partial, but complete or nearly
fo, muft have obliterated this original ftratification ; and the divifions they actually exhibit
can fcarcely be accounted for but on the fuppofition that they received them in the procefs by
which they confolidated. The thinnefs of thefe
bent layers too, in the greater number of rocks,
in micaceous fhiftus for example, favours this
opinion : it is fuch, that we cannot confider
the divifions of them as analogous to the divifions of beds or ftrata depofited in a horizontal
pofition from water, and afterwards elevated.
And the very forms of thefe contortions are too
complicated to admit of the fuppofition that
they have been produced by an expanfive power
applied to thefe rocks while foft.

The pofitions of the fecondary ftrata may be
afcribed partly to the fame caufe, and partly to
the direction of the bafe on which they recline.
It is certain, that cryftallization always com-
mences from the folid furface in contact with
the fluid ; to this the folid mafs adheres, taking
of courfe more or lefs perfectly its figure or pofition. In this manner is it conceived by Werner,
that the pofitions of the inclined ftrata have
been determined ; they have been depofited by

an imperfect cryftallization, mingled fometimes with a mechanical fubfidence, and have adhered to the fides of the primitive ftrata, on which they are incumbent. And, from the fame caufe, any bending which they have, different from that of the bafe on which they reft, may be explained. Hence we find, in conformity to this theory, that the fecondary ftrata are more inclined in the neighbourhood of the primitive vertical mountains; and, as they recede from thefe, are more nearly horizontal; and thefe alfo are more horizontal, in proportion as they are mechanical depofites.

Another circumftance may be pointed out, as having probably fome influence on the pofition of certain ftrata. It has been fuppofed, and the fuppofition is admitted by Profeffor Playfair not to be improbable, that, after having been depofited horizontally, fome ftrata may have funk unequally, from the unequal diftribution of their weight. A degree of inclination would thus be given to the whole, more or lefs, according to the degree of unequal depreffion that may have taken place. Werner has obferved, that fuch finkings of the ftrata at their formation are not improbable, and they fully account for the inclined pofitions of various ftrata which have not derived their arrangement from cryftallization.

Or if it should even be supposed, that the positions of the strata indicate the exertion of an expansive power from beneath, by which they have been elevated, such a cause, it may be remarked, may be admitted into the Neptunian system as not inconsistent with its principles. The general appearance of these strata prove, that, at the period of their formation, they had been subject to the most important revolutions. They have been disjoined from each other, the beds of rivers hollowed out, islands separated from the mainland; and, in a word, have suffered every species of fracture and dislocation. No cause can be assigned for these effects, more probable perhaps than that of an expansion produced from an accumulation of heat in the interior parts of the globe, the power of which has probably been spent in their production. No appearances in the elevation of these strata prove the peculiar tenets of the Huttonian theory, that the strata thus elevated had once been fused, or that a subterraneous heat has always acted, and still continues undiminished. On the contrary, it has been demonstrated, that these propositions are false, as being utterly inconsistent with the known laws of heat; and it will immediately be shown, that, from such a cause, the original horizontality of these strata cannot possibly be explained. Were it there-

fore clearly proved that they had been elevated by an expanfive power, we muft adopt
fome fuch fuppofition as that now ftated, that it
had been by a caufe acting at their formation,
either the accumulation of heat, the difengage-
ment of it by the procefs of cryftallization itfelf,
or the extrication of aerial fluids; a fuppofition
which fufficiently accords with the other principles of the Neptunian theory, and is probable in itfelf. One of this kind has even
occurred to a celebrated Neptunift, Sauffure.
He obferves, that after the materials were depofited, and formed horizontal beds, it may be
imagined that " heat, or elaftic fluids fhut up in
" the interior parts of the globe, had raifed up
" and broken the external cruft, had thus ex-
" truded the interior and primary part of that
" cruft, while the external and fecondary parts
" remained fupported on the others. Into the
" cavities formed by fuch eruptions the waters
" would fink, and leave the elevated land *."
Such a fyftem is fimilar to that firft fketched by
Lazarus Moro, and is a modification of the
Neptunian theory that may be maintained with
much advantage. It is doubtful, however, whether fuch an expanfive power is to be admitted
as a general caufe, or whether, if it has exifted
at all, it has not been local in its exertion, and
of courfe the caufe of particular appearances.

* Voyages aux Alpes, t. iv. 101.

I

We have next to confider the account which the Huttonian theory gives of the arrangement of the ftrata. In this firft, and perhaps moft important part of the proof from induction, it will be found fingularly defective.

It is fuppofed that the materials of thefe ftrata have been depofited from water in a loofe or divided ftate. From fuch a depofition it is not at all difficult to conceive that beds or layers, parallel to each other, might be formed : And if it were admitted that thefe materials could be confolidated merely from the exertion of the power of aggregation between their particles, their original arrangement might be preferved, and their fubfequent elevation by an expanfive power would produce the general appearances which the ftrata actually exhibit. But in the Huttonian fyftem it is maintained, that no confolidation can arife from fuch a caufe; it can be accomplifhed only by fufion, either partial or complete. Now, by fuch a fufion, it is obvious that the original divifion of the depofites into beds or ftrata, muft be completely fubverted; for each is rendered fo foft or fluid as to become homogeneous, and capable of yielding to preffure; and no caufe is pointed out how this fluid or foftened mafs is again to be formed into diftinct ftrata. This important objection may be illuftrated by a few examples.

Sandſtone is found in beds or ſtrata of very different degrees of thickneſs, generally horizontal in their poſition, or ſomewhat inclined. Theſe are ſuppoſed by the Huttonian geologiſt to have been formed from matter depoſited from the ocean, and ſoftened by ſubterraneous heat. But let him explain by what cauſe in their ſubſequent conſolidation they have been formed into diſtinct ſtrata. It may perhaps be ſaid to ariſe from the contraction of the maſs in cooling; and this appears to be the ſolution which the author of the theory gives : " The " contraction of the maſs conſolidated by fuſion, " or the effect of fire, is the cauſe of thoſe na- " tural diviſions in the ſtrata *." But why, it may be aſked, are the rents horizontal? the direction of all others the moſt unfavourable, from the gravity of the maſs ; and why do they preſerve this direction with ſo much regularity to ſo great an extent? Theſe circumſtances clearly prove that the parallelifm of theſe ſtrata is that which they originally had from their depoſition from water. Between the different ſtrata there are alſo often interpoſed thin layers of earth or clay, a proof that they had been formed ſucceſſively, and that on each of them, ſubſequent to its formation, this earthy matter had been

---

* Theory of the Earth, vol. ii. p. 56.

I 2

depofited ; that the whole had not been reduced into a foft or liquid mafs, which on cooling
had fplit into horizontal beds, but that the original ftratification had been preferved.

This is even admitted by Profeffor Playfair:
" In beds of fandftone nothing is more frequent
" than to fee the thin layers of fand feparated
" from one another by layers ftill finer of coal-
" ly or micaceous matter, that are almoft ex-
" actly parallel, and continue fo to a great ex-
" tent, without any fenfible diminution. Thefe
" planes can have acquired their parallelifm on-
" ly in confequence of the property of water,
" by which it renders the furfaces of the layers
" which it depofites, parallel to its own furface,
" and therefore parallel to one another *." But
if this be admitted, the difficulty occurs in its
original force. If the fandftone has been confolidated by fufion fubfequent to its arrangement, how have thefe divifions remained diftinct,
and preferved their exact parallelifm ? and how
has the interpofed clay efcaped confolidation?

Precifely the fame difficulty occurs in the
primitive ftrata, which, though now vertical,
have, according to the Huttonian geologift, been
originally horizontal. No particular fact could
be felected more forcible in illuftrating this dif-

* Illuftrations, &c. p. 44.

ficulty, than one ftated by Mr. Playfair: " In
" the micaceous and talcofe fhifti, thin layers
" of fand are often found interpofed between
" the layers of mica or talc. I have a fpecimen
" from the fummit of one of the higheft of the
" Grampian mountains, where the thin plates
" of a talcy or afbeftine fubftance are feparated
" by layers of a very fine quartzy fand not
" much confolidated*." Thefe kinds of fhifti
having more or lefs of a cryftalline texture,
muft have been nearly, if not completely, fufed,
as is indeed allowed, and even fuppofed. How
then have they remained divided into thin layers, the parallelifm of which is preferved with
more accuracy, and to a greater extent, than
even in the fecondary ftrata? Or how has the
fand, interpofed between thefe layers, efcaped
perfect confolidation? If it were equally fufible with the fhiftus, or nearly fo, it is obvious
that it ought to have been vitrified or aglutinat-
ed, and of courfe perfectly confolidated. If it
were lefs fufible, it is not lefs evident that it
ought to have been imbedded or cemented in
the fufed fhiftus. Thefe ftrata are fuppofed to
have been originally horizontal. Suppofe the
firft or undermoft ftratum confolidated, over this
has been depofited a layer of fand; the forma-

---

* Illuftrations, &c. p. 167.

I 3

tion of another ftratum of fhiftus by fufion, took
place above this : But it is evident that the
matter of this ftratum, fufed or foftened, muft
by its mere weight have enveloped the particles
of fand beneath it ; and when we reflect on the
fucceffion of thefe layers, the difficulty is much
increafed : the heat exerted on the undermoft
layers muft have been intenfe, and the preffure
upon thefe at the fame time great.   On no hypothefis, therefore, agreeable to the Huttonian
theory, could fand, loofe or little confolidated, be found between them, nor can the divifion of thefe layers themfelves be explained.

If we confider this difficulty with regard to
ftrata of different kinds alternating with each
other, the argument will be found to gain, if poffible, in force.   By what caufe have the divifions between thefe different ftrata been produced ? The fuppofition of fucceffive formation,
which in the Neptunian theory fully accounts
for fuch an arrangement, can in the Huttonian
have no place.   Let us fuppofe matter to be depofited, fufed or foftened, and confolidated, how
is another ftratum of a different kind of matter
to be formed above it ? Materials for it might
be collected, but thefe, according to the theory,
cannot be confolidated without heat. The heat
neceffary for this purpofe being applied from beneath, muft operate with ftill more force on the

ftratum on which thefe loofe materials lie, and
in innumerable cafes muft completely fufe it,
fince in the various pofitions and alternations
of the ftrata, the lefs fufible are often placed
over thofe which are more eafily fufed.   There
is no provifion therefore made in this fyftem for
the formation of a number of ftrata fuperincumbent on each other.   The upper ftratum
cannot have been laft formed, becaufe the heat
neceffary for its confolidation muft have fufed
thefe beneath, and the materials of it would of
courfe fink, and be imbedded in the fufed matter,
before they could be confolidated : The under
ftratum cannot poffibly have been formed, and
raifed fo as to be applied precifely to the one
above it, and this through an extenfive feries :
And laftly, the formation cannot be fuppofed to
be fimultaneous, for no caufe can be given why
by the neceffary degree of heat the whole fhould
not have been homologated, or how they fhould
afterwards have feparated into diftinct horizontal ftrata of the greateft extent, of different materials, and of various degrees of fufibility and
ftates of induration.

In examining the actual pofitions and connections of the ftrata, thefe difficulties might be
abundantly illuftrated; but to avoid a tedious
difcuffion, it may be fufficient to notice only a

I 4

few of thefe in which they are placed in the
cleareft light.

No alternation is more frequent that that of
calcareous and argillaceous ftrata. To account
for their confolidation, the Huttonian fuppofes
them to have been in a fufed or foftened ftate ;
but if they ever were in fuch a ftate, they muft,
from the attraction between the earths compofing them, have combined, and an entire homogeneous mafs, not a number of diftinct alternating ftrata, muft have been formed ; or at leaft
where they were contiguous, they muft have
united, and every trace of the original ftratification have been obliterated : yet fo far from
there being any union of this kind, the line of
feparation is perfectly diftinct, frequently fo
much fo that they are in firm adhefion.

Another alternation extremely common is
that of limeftone with argillite, the alternation
being continued through an extenfive feries.
One of thefe muft be lefs fufible than the other;
and as they alternate, it is a matter of perfect
indifference in the argument which of them is
fuppofed to be fo. Let it be fuppofed, as is probably the fact, that the limeftone is the leaft
fufible : A ftratum of it may be confidered as
commencing the feries : on it the materials of
the argillite have been depofited from the ocean,
and as thefe are fuppofed to be more fufible

than the limeſtone, it is poſſible that the central
heat may have operated through the latter,
without fuſing it, and have conſolidated the former. But above this ſtratum of argillite is another ſtratum of limeſtone. How could it have
been conſolidated ? By the ſuppoſition made, the
argillite is more fuſible ; the central heat could
not therefore operate through it, ſo as to conſolidate the materials of the ſuperior limeſtone
without fuſing it, and from this fuſion the materials of the limeſtone ought neceſſarily to have
been imbedded in the argillite. And when we
conſider that an alternation of this kind is often
found in an extenſive ſeries, by which both an
intenſe heat and a great preſſure muſt have been
exerted on the lower parts, it is obviouſly im-
poſſible that theſe different ſtrata could have
been conſolidated by a central heat, ſo as to be
kept diſtinct.

Strata of rock-ſalt are ſometimes covered by
ſtrata of ſandſtone, or limeſtone. The Huttonian geologiſt muſt ſuppoſe that this ſandſtone
has been conſolidated by the central heat acting
through the rock-ſalt below it. But this is
plainly an impoſſibility. The ſalt is a ſubſtance
comparatively very fuſible, as it can even be
volatilized by the heat of a coarſe pottery furnace, while ſandſtone is very infuſible. The heat
neceſſary therefore to ſoften ſandſtone, in this

pofition, muft have melted the falt beneath; and as this latter fubftance is of a much inferior fpecific gravity, the fandftone muft have funk in it, and the arrangement obferved in nature could never have been produced.

Laftly, we find in innumerable cafes ftrata more imperfectly confolidated than others above them, and of courfe farther removed from the confolidating power, though the difference cannot be afcribed to any difference in the fufibility of the fubftances compofing them. An example will place this in a clear light. In a fection of the ftrata at Newcaftle, coal is found at the depth of 102 feet. Over it is a bed of black clay 13 feet thick, with impreffions of ferns in its fubftance; above this, another bed of harder clay 26 feet in thicknefs. The ftratum incumbent on this is a hard quartzofe fandftone, with fpecks of mica, 25 feet thick; and this is again covered by clay. Now how could this fandftone have been confolidated by the fubterranean heat, while fo many feet of clay beneath it, and of courfe nearer the operation of that heat, had not even been indurated? We may pronounce it impoffible that it fhould be fo. Nor is the example uncommon: there are many fimilar to it, and even lefs favourable, as the banks of clay extend to 80, 100, or more fathoms, in thicknefs, with perfectly confolidated fandftone above;

and this is diverfified with alternations of limeftone, gypfum, coal, and a great variety of other
fecondary ftrata. It is not poffible to conceive
an arrangement which more clearly indicates
their origin, that thefe are fucceffive depofitions from water, varying in their confolidation, from the different forces of aggregation exerted between the particles of each ; and that
they have fuffered no other change to fubvert
or modify their original ftratification. Could
the Huttonian geologifts bring forward facts fo
forcible againft the Neptunian theory, it might
juftify the triumphant tone in which they have
fometimes conducted the controverfy.

The truth is, that the Huttonian theory has
in this part tacitly affumed the explanation of
the Neptunian fyftem,—that ftratification is the
effect of fucceffive depofition from water. Nor
has it been obferved, that the affumption is obvioufly incompatible with its principles; fince by
the fubfequent fufion, the original arrangement
muft have been deftroyed. The actual pofitions
of the ftrata are alfo totally irreconcilable with
the notion of their confolidation having been
effected by fufion, as is evident from the inftances pointed out ; the Huttonian fyftem therefore has the fingular infelicity of failing to account for the firft and moft important of all the
facts which are the fubject of a theory of the

earth,—the general ftratification of the materials at the furface of the globe.

It remains to confider the explanation given
in the Huttonian fyftem, of the appearances of
the UNSTRATIFIED ROCKS. Thefe are principally
granite, and the different varieties of porphyry,
bafalt, trap or whin, foffils, which, befides their
common character of in general not being ftratified, graduate into each other.

It has ufually been believed by geologifts,
that the principal rocks of this clafs are the moft
ancient in the globe. Granite, in particular, is
confidered as the bafe on which all others reft ;
and therefore as of prior formation to the others.
Dr. Hutton holds precifely the oppofite opinion,
that thefe rocks are of lateft formation, or pofterior to the ftrata, whether primary or fecondary, which are incumbent on them.

The formation of thefe rocks, in the Huttonian fyftem, is afcribed to the eruption of melted
matter from the internal parts of the earth, a-
mong the fuperincumbent ftrata. It is conceived, that, by the fubterraneous heat, which is always active, an immenfe quantity of matter is
kept in a melted ftate; that where the expanfive
power of this heat is increafed to a certain point,
the fuperincumbent ftrata are heaved up, and a
portion of this melted matter ejected ; and that

this, when confolidated by cooling, forms granite, porphyry, bafalt, or trap, according to the
compofition of the matter which has been in
fufion.

The arguments for the igneous origin of thefe
rocks, from their properties, are afterwards to be
confidered : At prefent we are to notice thofe
drawn from their pofitions.

The principal one of this kind, and on which
much is faid by the defenders of the Huttonian
theory, is that of veins of thefe unftratified
rocks, at leaft of whin and granite, running into
the neighbouring ftrata. It is admitted, that
veins are pofterior in formation to the ftrata in
which they are found ; and this appearance,
therefore, of veins of whin and granite entering thefe ftrata, prove, that the former have
been laft formed. It is fuppofed, that the matter of thefe had been injected, when in a fluid
ftate, into rents and cavities of thefe ftrata, and
when confolidated, had given rife to thefe appearances. And, as thefe veins often proceed
from immenfe maffes of whin or granite, or are
connected with them, it feems to follow, that
thefe muft have had the fame origin.

Thefe phenomena are explained by the Neptunift, from the fact, that granite is of different
formations. In the Wernerian theory, granite
is confidered as the rock to which the title of

Primitive moſt ſtrictly belongs, as being prior
in formation to gneiſs, micaceous ſhiſtus, and
others of the ſame claſs. But, although its poſitions with regard to theſe ſtrata generally indicate it to be ſo, yet in particular inſtances we are
compelled to admit that it has been of later formation. Thus it is ſometimes found, that a ſtratum of granite is incumbent on theſe other
rocks, or alternates with them, which proves
that it muſt have been poſterior to them in formation. Specimens have alſo been found of
granite, in which are contained pieces of gneiſs,
and even argillite, a demonſtration of the ſame
fact. This diverſity in the times of formation
of theſe rocks holds indeed with reſpect to them
all. When it is ſaid that granite is older than
gneiſs, and gneiſs than micaceous ſhiſtus, it is
not meant that it is univerſally ſo, but only generally, and not without exceptions, in each order of theſe rocks. This being eſtabliſhed, the
phenomena of granite veins penetrating the
other ſtrata, are eaſily explained; it belongs to
the claſs of facts now ſtated, which prove,
that, in ſome caſes, granite is of later formation than the other rocks named Primitive.
According to the Wernerian theory, veins were
originally fiſſures in rocks or ſtrata, which happened while theſe were ſtill covered with water
holding certain matters in ſolution, and were

filled by thefe matters being depofited or cryftallized. It is conceived, therefore, that, in this
manner, granite may have been introduced into
fiffures of the primitive ftrata.

Profeffor Playfair has obferved, that this diftinction between granite of early, and that of
recent formation, is purely hypothetical. " It
" is," fays he, " a fiction, contrived on purpofe
" to reconcile the fact here mentioned with
" the general fyftem of aqueous depofition, and
" has no fupport from any other phenomena *."
From the facts above ftated, it will be obvious,
however, that it is rather a direct inference from
phenomena; and the reafoning by which it is
deduced is ftrictly analogous to that by which
other geological conclufions are eftablifhed.
In all veins there are obferved undoubted marks
of different dates of formation, both with refpect to each other and to the ftrata; and there
are no grounds for fuppofing that granite veins
are an exception. The affumption, that the granite in mafs, and that in veins, are the produc-
tions of different periods, is fupported by other
facts than thofe from which it is deduced; thus,
it is obferved, that in many cafes, the granite in
veins is different in its properties from the other.
Nay, this point may be determined by an ap-
peal to the authority of Mr Playfair himfelf.
He admits, that whinftone, into which granite

* Illuftrations, &c. p. 320.

graduates, and which belongs to the fame clafs
of rocks, is of different formations: " Thefe un-
" ftratified rocks, diftinguifhed by the name of
" whinftone, are not all the work of the fame
" period; they differ evidently in the date of
" their formation ; and it is not unufual to find
" tabular maffes of one fpecies of whin inter-
" fected by another *." Now, as this is admitted
with refpect to whin, the probability of its being the fame with refpect to granite can fcarcely be denied, fince, in the Huttonian fyftem,
thefe two fpecies of rocks are believed to be
formed in precifely the fame manner. Not
only is there this prefumption, but the very
fact which is ftated by Profeffor Playfair, as
proving the different formations of whin, is true
with refpect alfo to granite. It has often been
obferved, that ftrata of granite are interfected
by veins likewife of granite, which are diftin-
guifhable, however, from the rock in which
they run, by being of a different colour or grain.
All veins are admitted, by every mineralogift,
to be pofterior in formation to the ftrata through
which they pafs ; and, from Mr. Playfair's own
reafoning, which, in this cafe, is perfectly juft,
this fact muft be admitted as decifive proof, that
thefe granites are not the production of the
fame period.

Granite, therefore, it may be confidered as

* Illuftrations, p. 81.

proved, is of different formations ; and the admiffion of this propofition affords a folution of
the phenomenon of granite veins, to which there
is no objection.

The fame folution may be given of the veins
of whin found in the ftrata ; as it is admitted to
have been formed at different periods, it is evident that thefe veins may be accounted for from
its being of recent formation.

It has been ftated as an argument in favour
of the igneous origin of thefe rocks, that where
they come in contact with the ftrata, or where
veins of them penetrate thefe ftrata, the latter
are indurated at the point of contact. This is
obferved particularly in whin, and it is confidered as a proof of the whin having been introduced in a melted ftate, and having, by its heat,
confolidated more perfectly the matter of thefe
ftrata. " Whether fandy or argillaceous, they
" are ufually extremely hard and confolidated ;
" the former, in particular, lofe their granulated
" texture, and are fometimes converted into
" perfect jafper *."

Thefe, and other alterations which the ftrata
fuffer, are afcribed, by Werner, to the action of
the folvent filling the vein. This will percolate
to a certain extent through the neighbouring
rock, and, by acting chemically upon it, as

* Illuftrations, &c. p. 73.

K

Werner fuppofes, or perhaps by depofiting part of the matter it has in folution, may alter its appearance and properties. Thus, in the prefent example, if, into a fiffure in a ftratum of fandftone, the fluid, holding the matter of trap in folution, were introduced, it would percolate to a greater or lefs extent through the fandftone, which is generally porous, and the matter depofited in the fandftone by this percolation might give it the hardnefs of jafper. Jafper is precifely fuch a foffil as might be expected to originate from an admixture of this kind, as it confifts of filiceous, with a large proportion of argillaceous, earth.

This explanation, as an hypothefis, is fufficiently probable, and the great proof that it is juft, is, that thefe alterations in the neighbouring rock are frequently fuch as the Huttonian theorift cannot explain, on the fuppofition of the matter of the vein having been in a fufed ftate, and having acted by its heat. They are not always induration, but often a change precifely the reverfe. Thus, when veins run through granite or gneifs, the felfpar is often changed into kaolin or fine clay, as it is by expofure to air and moifture; the mica is alfo decompofed, and in fiennite the hornblende, fo that in the latter a green friable earthy matter is produced. " The mountain of Scharfenberg is of red gra-

" nite, very hard, and of an equal grain; it is
" traverfed by metallic veins, of which the *gan-*
" *gue* is white or reddifh quartz, calcareous fpar,
" and white clay. The granite is commonly
" much decompofed along the fides of the veins;
" neither its felfpar nor mica are to be found;
" thefe two fubftances are fupplied by a grey
" clay, and a green earth, of the nature of ftea-
" tite : this decompofition extends ten or
" twelve inches on each fide of the vein; be-
" yond this the granite refumes its ufual ap-
" pearance. Sometimes a fhiftofe rock is chan-
" ged into a foft clay, which forms a border
" fome inches thick along the fides of the
" vein *."

It is obvious, that thefe are changes precifely
oppofite to what would have been produced by
heat, while they are fuch as might have originated from the percolation of a fluid, and are
thofe, indeed, which are produced by expofure
to moifture. If we fhould, therefore, refer the
change produced by veins of whin on the ftrata
of fandftone to a different caufe, we fhould at
once lofe all unity of theory, and relinquifh a
principle capable of affording a fatisfactory explanation. The appearance in fandftone is one
which may juftly be afcribed to the fame caufe
that has produced the change in the granite or

* Journal des Mines, No. xviii. 85.

K 2

gneifs; and its particular nature may be afcribed
to the peculiar action of the percolating fluid,
and the nature of the fandftone itfelf. The contraft between the two hypothefes is evident: the
Wernerian explains from one principle *all* the alterations in the rocks contiguous to veins; the
Huttonian explains only *one fpecies* of alteration,
while there are others not only inexplicable, but
inconfiftent with the fuppofition by which that
one is folved.

In fome cafes alfo it appears that the formation and filling of veins have been nearly fimultaneous with the confolidation of the ftrata; and
in fuch cafes, Werner has remarked that the
gangue participates confiderably of the nature
of the rock, and the rock in its turn is more
impregnated with the matter of the vein. If
this has been the cafe with fome of the veins of
whin in the ftrata, the intermixture, and the
induration arifing from it, might be more complete.

The induration which is obferved where beds
of whin come in contact with ftrata of fandftone,
may be explained on the fame principle; either
from the percolation of the fluid loaded with
particles of whin, or from a degree of intermixture from a formation nearly fimultaneous. The
fame explanation may be given of the induration of fragments of fandftone fometimes found

in veins of whin, as thefe are in a fituation in which this percolation muft have taken place to the greateft extent.

Another fact may be added, which proves that this induration muft in all thefe cafes be afcribed to fome fuch caufe : It is, that fandftone is often highly indurated when in contact with fubftances where the caufe affigned by the Huttonian geologift,—the application of a ftrong heat, could not have acted. Sandftone much indurated, it is obferved by Mr. Jamefon, is often found covered by clay and wacken*: and it will not be affirmed that clay, by any heat it ever had, could caufe the induration of the ftratum of fandftone on which it is incumbent, while by the operation pointed out in the preceding explanation, fuch an effect may have been produced.

Dr. Hutton obferved that where veins of whin run into ftrata of coal, the coal contiguous to it is frequently found changed in its properties. " It has loft its fufibility, and appears to be re- " duced nearly to the ftate of coke or charcoal:" a change which is afcribed to the melted whin having by its heat expelled the bituminous matter of the coal. The fact does not always correfpond with this obfervation, but is fometimes

* Nicholfon's Journal, vol. iii.

K 3

the reverfe. But where it does occur, it may, in common with other alterations of the ftrata by veins running through them, be afcribed to the operation of the fluid by which the vein was filled. This percolating through the coal might be capable of producing changes in its compofition and properties; it might depofite a portion of earthy matter which would render the coal lefs inflammable: and when we confider that a fimilar caufe can occafion the decompofition of felfpar, mica, and hornblende, there is no improbability in the fuppofition that it might carry off, or caufe the decompofition of the bituminous part of the coal. To either or both caufes may be owing the change in the properties of that foffil.

This fact is important in another point of view, as it may probably lead to a demonftration of the falfity of the Huttonian hypothefis. The difference between whin and lava is faid to be, that the former has been in fufion under an immenfe compreffion, while the latter has been in fufion at the furface; and this difference is ftated as the caufe of fome peculiarities in the one compared with the other. It is given as the caufe of the lava being more or lefs cellular or porous, while the whin is folid, as in the latter the air had been retained, which in the former had efcaped; and likewife as the caufe of the

whin containing calcareous fpar, while the lava
does not, the compreffion prefent in the fufion
of the former having prevented the expulfion
of the carbonic acid. It is admitted, therefore,
that wherever whin has been in fufion without
this compreffion, it ought to affume the precife
characters of lava ; it ought to be more or lefs
porous, and ought to contain no carbonate of
lime. Now in the fact above ftated, with refpect to the coal, it is affumed that compreffion
had not been prefent, as the coal had allowed
its bituminous matter to be expelled. It will
not be contended that compreffion could be removed from the coal, and prefent on the melted
whin, at the diftance of a few inches. If therefore any carbonate of lime fhould be found in
the whin in fuch a fituation, either in its compofition, or in a feparate cryftallized ftate, (and
the probability is, that it will be found ;) or even
if the whin fhould not differ fomewhat in its appearance from common whin, and approach to
lava, we fhall obtain unanfwerable demonftration of the falfity of fome of thofe fuppofitions
which conftitute the Huttonian theory.

Again it is ftated, that " the difturbance of
" the ftrata, wherever veins of whinftone a-
" bound, if not a direct proof of the original
" fluidity of the whinftone, is a clear indication

K 4

" of the violence with which it was introduced
" into its place *."

But that fuch appearances are not effects neceffarily produced from the invafion of the ftrata
by ftreams of melted whin, and do therefore not
indicate fuch an invafion, is proved by there being frequently no figns of difturbance whatever
accompanying veins either of this rock or of
granite. Veins of whin traverfe coal without
any diflocation of the bed of coal; and veins
of granite are often unaccompanied by thefe
marks of violence which might be expected
from its irruption into the neighbouring ftrata.
The abfence of thefe in many cafes affords an
argument at leaft equivalent in force to their
prefence in others; and indeed, even where
they do occur, it is admitted that they do not
prove the original fluidity of thefe maffes.

In the Neptunian fyftem there is no difficulty
in explaining all thefe appearances, and fome of
them are even ftated by Werner as proofs of
his theory of veins. According to that theory,
the fiffures in rocks and ftrata, which, when fub-
fequently filled, form veins, have in general
arifen from an unequal fubfidence of the rock
in which they are found, one part of it from a
difference in its height, in the materials compofing it, or in the want of fupport from contiguous
rocks having funk more than the other. It is

* Illuftrations, &c. p. 74.

obvious, therefore, that the rock on the one fide, of the fiffure produced by fuch a caufe, will be lower than the rock on the other, and hence the fhifting of the ftrata obferved by the fides of veins, Thefe finkings of the ftrata, producing fuch fiffures, have alfo taken place at various times, as is proved by feveral facts in the appearances of veins, particularly by that of one vein croffing another; and from this circumftance the difturbances fometimes prefent are eafily accounted for. If after a vein was formed, a new fubfidence had taken place in the rock in which it was fituated, it is evident that diflocation, both of the vein itfelf, and of the contiguous parts, muft have taken place to a greater or lefs extent.

It is to be obferved alfo, as a fact highly favourable to the Neptunian theory of thefe difturbances, that the fiffures where thefe flips or heavings of the ftrata take place, are frequently filled with fubftances, fuch as foft clay or fand, which cannot be fuppofed to have been ejected with any projectile force, or to have produced any alteration of the exifting arrangement by the violence of their introduction.

It is further ftated, that when whin is found interpofed in tabular maffes between beds of ftratified rocks, " it is not uncommon to find " the ftrata in fome places contiguous to the

" whin, elevated, and bent, with their concavity
" upward, fo that they appear clearly to have
" been acted on by a force that proceeded from
" below, at the fame time they were foftened,
" and rendered in fome degree flexible *." An
appearance of this kind, however, might equally
be produced from the fubfidence of certain parts
of the ftrata while they were foft and yielding,
as from the elevation of other parts of them;
and confequently, fuch an appearance affords no
certain indication of the nature of the caufe that
has operated on them.

The laft argument, from the pofition of whin,
is thus ftated by Profeffor Playfair: " If it be true,
" that the maffes of whin thus interpofed among
" the ftrata, were introduced there after the
" formation of the latter, we might expect to
" find, at leaft in many inftances, that the beds
" on which the whinftone refts, and thofe by
" which it is covered, are exactly alike. If
" thefe beds were once contiguous, and have
" been only heaved up, and feparated, by the
" irruption of a fluid mafs of fubterraneous la-
" va, their identity fhould ftill be recognifed.
" Now, this is piecifely what is obferved; it is
" known to hold in a vaft number of inftances,
" and is ftrikingly exemplified in the rock of
" Salifbury Craig, near Edinburgh.

* Illuftrations, &c. p. 75.

" The fimilarity of the ftrata that cover the
" maffes of whinftone, to thefe that ferve as the
" bafe on which they reft, and again, the diffi-
" milarity of both to the interpofed mafs, are
" facts which I think can hardly receive any
" explanation on the principles of the Neptu-
" nian theory. If thefe rocks, both ftratified
" and unftratified, are to be regarded as pro-
" ductions of the fea, the circumftances would
" require to be pointed out which have deter-
" mined the whinftone, and the beds that are
" all around it, to be fo extremely unlike in
" their ftructure, though formed at the fame
" time, and in the immediate vicinity of one
" another ; as alfo thofe circumftances on the
" other hand, which determined the ftratified de-
" pofites, above and below the whinftone, to be
" precifely the fame, though the times of their
" formation muft have been very different. The
" homogeneous fubftances thus placed at a di-
" ftance, and the heterogeneous brought fo
" clofely together, are phenomena equally un-
" accountable, in a theory that afcribes their
" origin to the operation of the fame element,
" and that neceffarily dates their formation ac-
" cording to the order in which they lie one
" above another *."

* Illuftrations, &c. p. 75.

The argument is here ftrongly ftated; but nothing is more eafy than to fhow that it is equally conclufive againft the Huttonian theory; that the difficulty which is urged is a particular example of one more general, which is common to both. In the view which is prefented to us, our attention is confined to the alternation of whin with the ftratified rocks; but all that is faid applies with the fame force to the alternation of the ftrata with each other. Thefe, according to both theories, are formed in the fame manner; and the general difficulty prefents itfelf, how we fhall account for a ftratum of a particular kind being formed, for the formation of it being interrupted, while one of a different kind is produced above it; and for the formation of the firft being again refumed, fo that it fhall be preoifely fimilar to the one beneath, though it is cut off from all communication with it, by the ftratum of a different kind which is interpofed. This will be rendered evident by an illuftration: Among the vertical ftrata, micaceous fhiftus and flate frequently alternate with each other; according to the Huttonian theory, thefe, previous to their elevation, were in a horizontal pofition; the micaceous fhiftus, fuppofe, is undermoft, and muft have therefore been firft formed; over this is the flate; and again, above this, is the micaceous shistus, the same as beneath. In this case, it cannot be supposed that the slate has been introduced in fusion between the two strata of shistus, for there is frequently an extensive series of alternations, and both of these rocks are confidered, even in the Huttonian system, as stratified, and of the same formation. We have here, therefore, the difficulty which Professor Playfair so strongly urges, of accounting for the similarity of strata between which another is interposed, and the dissimilitude of those to this interposed mass. In the strata which remain horizontal, the case is precisely the same. An alternation which not unfrequently occurs, is that of limestone with argillite. The strata of limestone, between which the argillite is interposed, are alike in their properties; yet it will not be supposed that these strata have been divided, by the argillite having been introduced between them in fusion. They must be admitted to be of successive formation; and the argument of the learned Professor may, in this case, be directed with its full force against his own system; for, if they have been formed in the same manner, as is admitted, what cause can have determined the formation and deposition of the argillite, between the formation of the strata of limestone? and how should these latter be alike, when the times of their formation muft neceffarily have been different?

The force of this objection can now be appreciated. If the fimilarity of ftrata, feparated by interpofed maffes of another kind of rock, were confined to that particular cafe where whin is the interpofed fubftance, the argument thence derived in favour of the Huttonian fyftem would be of fome force : But the phenomena is much more general; it is equally obfervable in the alternations of ftrata, whether primitive or fecondary, with each other; and the Huttonian fyftem can in fuch cafes afford no explanation more fatisfactory than the Neptunian, of the fimilarity exifting between the feparated ftrata, and the diffimilarity of thefe to the ftratum interpofed between them. It muft not be forgotten that thofe inductions, which conftitute our theories, cannot in every point be complete; and appearances muft frequently be found which, from our imperfect knowledge, cannot be fully explained. The prefent is one of this kind. Whether we afcribe the formation of the ftrata to the fole agency of water, or to the operation of fire, we fhall find it equally difficult to affign precife caufes for the total difference in nature of contiguous foffils and rocks, which muft have been formed at nearly the fame period, or for the feries of alternations which they obferve with

respect to each other. Werner considers such
facts as proofs that the original or chaotic fluid
had, at different periods, held different substances
in solution, and that from this had originated
the successive strata of different kinds. The
same solution too might successively afford de-
posites of different kinds, according to the affinities that had been exerted having been varied
by the very combinations taking place.

This alternation of the strata leads to the consideration of a point of much importance in geology,—the transition of different kinds of rock
and strata into each other. In those examples
of alternation, which have been the subject of
the argument now discussed, the line of separation between the contiguous masses is supposed
to be perfectly distinct, as it frequently is ; but
in many other cases the transition is gradual, or
the one kind of rock passes insensibly into the
other; and where this transition takes place
from rocks of the unstratified into those of the
stratified kind, it is a phenomenon which overturns the Huttonian hypothesis of the formation
of these rocks.

According to that hypothesis, the stratified
and unstratified rocks have been formed at very
different periods, and even in different modes.
The former have been first arranged and consolidated, and afterwards the matter composing

the latter has been thrown up in a fufed ftate, and has invaded the others, filling up every cavity. According to this view of the fubject, it is evident that the line of diftinction between them fhould always be well marked. A flight alteration might take place in the appearance or hardnefs of the ftratified rock, from the heat of the melted matter brought into contact with it ; but there fhould be no fuch thing as an imperceptible gradation or tranfition of the one into the other, while, if both are formed in the fame manner as the Neptunian theory fuppofes, fuch a gradation is not more than what might be expected occafionally to take place.

This tranfition, therefore, of the unftratified into the ftratified rocks being fo unfavourable to the Huttonian hypothefis, is attempted to be denied : " If, in thefe inftances, (fays Mr.
" Playfair) the gradation were infenfible, as fome
" have afferted it to be, between the ftrata and
" the interpofed mafs, fo that it was impoffible to
" point out the line where the one ended and the
" other began, whatever difficulties we might
" perceive in the Neptunian theory, we fhould
" find it hard to fubftitute a better in its room.
" But the truth feems to be, that, in the cafes we
" are now treating of, no fuch gradation exifts;
" and that, though where the two kinds of rock
" come into contact a change is often obferved,

" by the ftrata having acquired an additional de-
" gree of induration, yet the line of feparation
" is well defined, and can be precifely afcertain-
" ed. This is at leaft certain, that innumerable
" fpecimens, exhibiting fuch lines of feparation,
" are to be met with; and wherever care has
" been taken to obtain a frefh fracture of the
" ftone, and to remove the effects of accidental
" caufes even where the two rocks are moft
" firmly united, and moft clofely affimilated, I
" am perfuaded that no uncertainty has ever re-
" mained as to the line of their feparation. For
" thefe reafons, it feems probable that the gra-
" dual tranfition of bafaltes into the adjoining
" ftrata, is in all cafes imaginary, and is, in
" truth, a mere illufion, proceeding from hafty
" and inaccurate obfervation *."

The obfervations, however, by which the
tranfition of the unftratified into the ftratified
rocks are eftablifhed, reft on authorities too unexceptionable to be overturned by an affertion
of this kind. They are eftablifhed on the authority of Werner, allowed to be the moft fkilful mineralogift of his age, and whofe obfervations, becaufe they run counter to the views
of the theorift, are not to be termed hafty and
inaccurate, or mere illufions; and they are fup-

* Illuftrations, &c. p. 44.

L

ported by the authorities of Sauffure, Charpen-
tier, Reufs, and other geologifts of the higheft
character. Profeffor Playfair indeed, in a great
meafure, recalls, in a fubfequent part of his
work, the affertion that has been quoted. " I
" am difpofed," fays he, " to make fome limita-
" tion to what is faid in § 72, where I have ex-
" preffed an abfolute incredulity as to fuch tran-
" fitions as are here referred to. The great fkill
" and experience of the mineralogift who has
" defcribed the ftrata at Scheibenberg, do not
" allow us to doubt of his exactnefs, though
" fome of the appearances are fuch as decom-
" pofition and wearing might well enough be
" allowed be fuppofed to produce." The manner in which this acknowledgment is expreffed
fufficiently fhows the reluctance with which the
fact is conceded. It is indeed one which bears
with the utmoft force againft the Huttonian hypothefis of the unftratified rocks, and which is
but feebly obviated by an improbable fuppofition.

The obfervations of Werner eftablifhing this
tranfition of bafalt into ftratified rocks, are of
the firft importance, and his reafoning from them
convincing and juft. At the bottom of the hill
of Scheibenberg, he obferves, " there is firft a
" thick bank of quartzy fand ; above that, a bed
" of clay ; then a bed of the argillaceous ftone

" named wacken ; and upon this lay the bafalt;
" When I faw the three firft beds running almoft
" horizontally under the bafalt, and forming its
" bafe, the fand becoming more fine above ;
" then argillaceous, and at length changing in-
" to real clay ; as the clay was converted into
" wacken in the inferior part ; and finally the
" wacken into bafalt ;—in a word, when I found
" a perfect tranfition of pure fand into argilla-
" cious fand, of this into fandy clay, and of the
" fandy clay, through many gradations, into fat
" clay, wacken, and laftly bafalt,—I was irre-
" fiftibly led to conclude, (as every impartial
" judge, ftruck with the confequences of that
" phenomenon, would have been) that the
" bafalt, wacken, clay, and fand, are of one
" and the fame formation ; they are all the ef-
" fect of a precipitation by the humid way, dur-
" ing one and the fame fubmerfion of this coun-
" try ; the waters which covered it brought at
" firft the fand, afterwards depofited the argil,
" and changed gradually their precipitation in-
" to wacken, and laftly into true bafalt."

After again ftating fome remarks, to render doubtful, if poffible, the obfervations of
Werner on thefe tranfitions, which fhew the
unwillingnefs with which they are admitted,
Profeffor Playfair propofes an hypothefis, to account for them on the Huttonian fyftem : " It

" is certain," fays he, " that the bafis of whin-
" ftone, or the material out of which it is pre-
" pared by the action of fubterraneous heat, is
" clay in fome ftate or other, and probably in
" that of argillaceous fhiftus. It follows of
" confequencc, that argillaceous fhiftus may,
" by heat, be converted into whinftone. When,
" therefore, melted whinftone has been poured
" over a rock of fuch fhiftus, it may, by its
" heat, have converted a part of that rock in-
" to a ftone fimilar to itfelf; and thus may
" now feem to be united, by an infenfible gra-
" dation, with the ftratum on which it is in-
" cumbent; and phenomena of this kind may
" be expected to have had really happened,
" though but rarely, as a partiular combina-
" tion of circumftances, feems neceffary to pro-
" duce them*."

The conclufions here follow each other with
great rapidity, as if they were obvious and undeniable ; yet the whole is a feries of hypothefes
brought forward to reconcile, if poffible, the
fact of the gradation of thefe rocks with the
Huttonian fyftem, but unfupported by any
proofs, and even improbable in themfelves. It
is not proved, or rendered probable, that whin
is formed by fire from argillaceous fhiftus. On

* Illuftrations, p. 285.

the contrary, the compofition of bafalt is peculiar, as according to the analyfis of it by Dr. Kennedy, it contains a quantity of foda, which is
comparatively a rare ingredient in the mineral
kingdom. Though it were proved, it is not argillaceous fhiftus, but unconfolidated clay that
the bafalt could have come into contact with
in thefe ftrata; and let Profeffor Playfair endeavour to eftablifh his hypothefis, by converting
clay into bafalt by the application of heat.
Nor, laftly, if the circumftances were of the
moft favourable kind; if even the ftratum with
which the bafalt was in contact, was actually
the material from which bafalt could be
formed; the heat of the ftratum of fufed bafalt,
fuppofed to be introduced, could not be fufficient to convert a large bed of it into wacken.
We know that the heat neceffary to fufe bafalt
is very confiderable; we obferve, that at the
junctions of whin with the ftrata, there is often
fcarcely any change, and where any is percept-
ible, it does not extend above a few inches.
What circumftances, therefore, could have been
prefent, which enabled this ftream of fufed
bafalt to convert, by its heat, a large bed of clay
into wacken, into which it imperceptibly gra-
duated. If bafalt, in a ftate of fufion, is capable,
by its heat, of converting clay into a fubftance
analagous to itfelf, it is obvious that the ftrata

L 3

which are in contact with it, should always be materially changed; and, above all, if this hypothesis be true, no such arrangement could possibly exist, as a bed of unaltered clay or shistus beneath, and in contact with a stratum of basalt. Yet such connections of these strata are not unfrequent. Bergman, for instance, states an example of basalt incumbent on thin beds of clay or bituminous shistus; and Kirwan, and other mineralogists, mention instances of trap alternating with argilite. The Huttonian, therefore, is reduced to the dilemma of failing to account, either for the gradation of clay into basalt, or the contact of this rock with unaltered clay. Lastly, the general gradation observed in this mountain is overlooked. It is not confined merely to that of the clay into the wacken, but extends from the sand to the basalt, through the whole series; and it is transgressing obviously the rules of just reasoning, to confine the explanation to one part of it merely, and ascribe this to a local cause, or to refer that part to one cause, and the rest to another.

Neither is this the only example of the gradation of unstratified into stratified rocks. It is also observed occasionally in granite; and is indeed a geological truth, of which there can be no doubt. In the Neptunian system, as both these classes of rocks are supposed to be of similar formation, the phenomenon is not furprifing, but, in the Huttonian hypothefis, it cannot be explained. Their formation is ftated to be entirely different. The matter of the unftratified mafs is fuppofed to be thrown up in a fufed ftate among the ftrata; it is applied to them in this ftate, and may therefore be united to them, but it is in direct contradiction to fuch a fyftem, that the one fhould infenfibly graduate into the other. In every hypothefis, a number of the phenomena to which it relates, will be apparently explained; but if its bafis is not in truth, there will always be fome which cannot be brought under it, but ftand in oppofition to its affumptions, proving their fallacy. Such a fact, with regard to the prefent hypothefis, is the gradation of the unftratified into the ftratified rocks; and fo clear are the indications it affords of the origin of thefe maffes, that it is with juftice it has been confidered as one of the ftrongeft proofs of the Wernerian fyftem. It gave a deadly blow to the theory of the Vulcanifts, and it inflicts one not lefs fatal to the Plutonic hypothefis.

Another fact fcarcely lefs unfavourable to the Huttonian hypothefis of the formation of thefe rocks is, that they are not unfrequently, both granite and trap, found ftratified. The ftratification of granite, both horizontal and verti-

L 4

cal, is eſtabliſhed by the obſervations of Sauſſure, and other mineralogiſts, and is admitted
by Mr. Playfair; yet, as granite, according to
the ſyſtem he defends, conſiſts of matter in perfect fuſion, thrown up from the central regions,
we do not ſee how it could have formed any o-
ther than an irregular maſs, and we might with
as little reaſon expect to find ſtratification in it
as in a bank of lava.

This difficulty is attempted to be removed by
the following explanation. " Rocks, of which
" the parts are highly cryſtallized, are already
" admitted as belonging to the ſtrata, and are
" exemplified in marble, gneiſs, and veined gra-
" nite. In the two laſt we have not only ſtra-
" tification, but a ſhiſtoſe, united with a cryſ-
" tallized ſtructure, and the effects of depoſition
" by water, and of fluidity by fire, are certain-
" ly no where more ſingularly combined. The
" ſtratification of theſe ſubſtances is, therefore,
" more extraordinary than even that of the moſt
" highly cryſtallized granite. Neither the one
" nor the other can be explained, but by ſup-
" poſing, that while ſuch a degree of fluidity
" was produced by heat, as enabled the body
" when it cooled to cryſtallize, the whole maſs
" was kept in its place by great preſſure, act-
" ing on all ſides, ſo that the ſhape was preſerv-
" ed as originally given to it by the ſea *."

* Illuſtrations, p. 336.

It is not eafy to difcover what is here meant
by the fhape originally given by the fea being
preferved. Granite is fuppofed, in the Huttonian fyftem, to be matter which has been completely fufed in the central regions, and erupted ; and if it has undergone thefe operations of
perfect fufion, and of eruption, it cannot furely
be imagined that it could have preferved its original ftratification. And if it is fuppofed, that
ftratified granite may not have been formed in
this manner, but that its materials have been
fufed in the place where they were depofited,
the unity of the theory is entirely loft, and two
hypothefes refpecting the origin of this rock are
actually advanced.

It is befides, impoffible on any fuppofition, to
believe that matter completely in fufion, let it
be fubjected to what preffure it may, could preferve its original divifion into beds. That preffure would not prevent the gravity of the upper
parts of the mafs from being exerted on the under, and if it was fluid, this preffure muft have
obliterated every trace of feparation. The ftratification of granite, therefore, cannot be explained by the hypothefis Profeffor Playfair has
advanced.

The ftratification of trap furnifhes an objection not lefs conclufive, and perhaps more ftriking, as its ftrata are generally alternated with

others. Of this a ftriking example has been
ftated as exifting at Habichtfwalde, near Caffel:
" On fecondary limeftone are found ftrata of
" fand, clay, wacken, and bafalt, which alter-
" nate with each other not lefs than three times,
" and always in the fame order. Over the third
" ftratum of bafalt is found a thick bed of coal,
" which is covered by a quartzofe fandftone,
" containing the remains of plants, and petrified
" wood. Laftly, on this fandftone are found,
" in the fame order as before, ftrata of clay,
" wacken, and bafalt; in the fand are found
" marine fhells ; the ftrata of bafalt have a baf-
" altic *tufa* intermixed with them, containing
" fragments of bafalt, olivin, and vegetable re-
" mains *." Reufs and Dolomieu have alfo obferved ftrata of bafalt, alternating with ftrata of
limeftone.

The argument which this difpofition of trap
furnifhes againft the Huttonian theory, of its
formation, is very evident. It is fuppofed to
have been thrown up among the ftrata, and to
have filled any cavities or fpaces between them.
In this manner, it might be conceived, that a
rock, or mountain of trap, might be formed.
But, is it poffible that it could have formed ftrata alternating with others? What are the fuppofitions neceffary to account for fuch an ar-

* Traite. de Mineralogie par Brochant, tom. ii. p. 609.

rangement as that ftated above ?—that a ftratum
of limeftone fhould have exifted, then a bed of
clay, another of fand, and over this an empty
fpace, correfponding in fize and direction to the
ftrata beneath; that this arrangement fhould be
repeated three times; and that thefe empty
fpaces fhould have afterwards been filled by fufed bafalt and wacken injected into them. It
may at once be affirmed, that fuch fuppofitions
are inadmiffible; and that were they made, their
extravagance would afford a fufficient refutation
of the fyftem that required them. If it fhould
be faid, that, in fuch cafes where trap is ftratified, it is formed in the fame manner as the other
ftrata, by its materials having been depofited,
and foftened or fufed; this is at leaft relinquifh-
ing the fimplicity of the theory, and propofing
two modes of formation of this rock. And
even this facrifice would prove inadequate, for
the above arrangement in which *fand* and *clay*
alternate with the others, completely excludes the
operation of fire, and proves that the whole muft
have been formed by the agency of water.

Laftly, the remains and impreffions of organic
fubftances found in trap demonftrate its aqueous origin. In this country, in which this rock
is abundant, fuch appearances are undoubtedly
rare, and indeed have not been obferved. But
in other countries, they appear to be far from

being uncommon. Werner found in a vein of wacken, at a depth of not lefs than 150 fathoms, trees, with the branches, and even leaves, petrified ; and it is ftated, on the fame authority, that different rocks of the trap kind, contain marine fhells, and even bones of quadrupeds. Nothing can be more obvious, then that fuch fubftances could never have been contained in a ftream of melted matter thrown from the central regions : or if they had even fallen into it after its irruption, and while ftill fluid, as it is fuppofed may have been the cafe, they muft have been deftroyed by the fufed matter ; and indeed the circumftances, in fome of thefe cafes, are irreconcilable with fuch a fuppofition.

From this review of the Huttonian theory of the unftratified rocks, it muft be evident, that it is attempted to be fupported by appearances which admit of an equal, and in fome cafes a more fatisfactory explanation, from the oppofite opinion. The occafional ftratification of thefe rocks, their gradations into thofe which are ftratified, and their alternations with other ftrata, are inconfiftent with that opinion, and indubitably prove their aqueous origin.

———

Under this clafs of arguments, may be confidered thofe drawn from the appearances of veins, the fubftances filling their cavities falling

[157]

properly under the defcription of unftratified minerals.

Veins are accurately defined by Profeffor Playfair, " feparations in the continuity of a " rock, of a determinate width, but extending " indefinitely in length and depth, and filled " with mineral fubftances different from the " rock itfelf*." It is admitted by mineralo-gifts, that thefe veins are fubfequent in forma-tion to the rocks or ftrata in which they are found ; and the cryftallized ftate of the fub-ftances with which they are commonly filled demonftrates that thefe had been introduced in a fluid form.

The Huttonian theory of veins is, that they have been formed by injection, that the matter filling them has been thrown up in a ftate of fufion, and that this fufed matter entering the rents and cavities of the ftrata, has confolidated and produced mineral veins.

According to the Wernerian theory, the ca-vities of veins have been originally fiffures in the ftrata or rocks, produced while thefe were yet foft and covered by the waters, by various caufes, principally by the unequal finkings of thefe maffes at their confolidation, occafioned by the various denfities of the fubftances com-

* Illuftrations, p. 57.

pofing them, and their unequal elevations, and
by the diminution of the waters by which the
mechanical fupport afforded to the mountain
at its fides was withdrawn. The appearances
of veins are precifely fimilar to what might be
expected to be the appearances of fiffures produced from fuch caufes, as, in their taking a direction
approaching more to the vertical than horizontal,
in their proceeding without much inflection, in
the diameter of them diminifhing after they have
proceeded a certain extent, and in their at length
difappearing entirely, by the approach of their
fides, while they generally continue open to the
furface, though of a diminifhed fize.

Into the fiffures arifing from thefe caufes, the
fluid ftill holding much matter diffolved, and covering the furface, would of courfe find accefs;
and the folution in thefe cavities being at reft, the
cryftallizations from it of the diffolved matter,
would take place with more regularity, and they
would be more completely feparated from each
other, than in the formation of the ftrata. Many
of thofe veins have alfo been filled by fucceffive
cryftallizations, and hence a greater variety of
fubftances have been introduced into them.

The formation of thefe veins is of very different dates, the fiffures from which they arife
having happened at different times. This is
evident, from the fimple fact, that one vein of-

ten croffes another without interruption or alteration of the matter it contains. A new fiffure has taken place traverfing the exifting vein,
and has been afterwards filled, forming a new
vein, and this, it has been proved by repeated
croffings, has taken place a number of times
fucceffively. Such repeated fiffures often de-
range the direction of the vein, and its relative
fituation with refpect to the fides of the rock in
which it exifts.

Several of the arguments in fupport of the
Huttonian theory of mineral veins are drawn
from the nature and properties of the fubftances
they contain, fuch as their infolubility in any
one menftruum, and the mutual impreffions in
their cryftallization. Thefe are afterwards to be
noticed ; at prefent we are to confider thofe deduced from the ftructure and pofitions of the
veins.

That the matter filling the veins has not been
introduced in a ftate of folution, is evident, it is
alleged, from their being no trace of that folvent in the vein, and from the vein itfelf being
completely filled up.

But, if the entrance into the vein has been
open, as is maintained, thefe appearances are not
different from what might be expected. Confolidation, or cryftallization, it has already been
fhown, may take place from folution, without

any fenfible portion of the folvent being retained ; and the matter, confolidating in an open
cavity, would exclude the fluid feparated, while
if a frefh quantity of the folution had accefs, the
cavity might be completely filled up. Or, if
the fluid was left, it would penetrate the furrounding matter by infiltration, and leave the
vein only partially filled.

It is faid, " if the veins were filled by depo-
" fition from above, we ought to difcover in
" them fuch horizontal ftratification as is the
" effect of depofition from water." The ufual
ftructure of a mineral vein is that of incruftations, or parallel coats, on its fides. One fubftance, calcareous fpar, for example, or quartz,
adheres immediately to the fides of the rock,
and next to this a mafs of any metallic ore, and
thefe may be varioufly intermixed with each o-
ther, and even with other fubftances. But, if
the Neptunian theory were juft, it is affirmed,
that the materials fhould be difpofed in horizontal layers acrofs the vein, inftead of being parallel to its fides. " On no fuppofition," it is faid,
" can thefe incruftations be received as a proof
" of aqueous depofition : It may, indeed, be
" certainly inferred from them, that the matter
" which they confift of, was fluid at the time of
" their formation ; but the abfence of all ap-
" pearance of horizontal difpofition in any part

" of the vein, amounts nearly to a demonftra-
" tion, that this fluidity did not proceed from
" folution in a menftruum." And again, " if,
" as the Neptunifts maintain, the materials in
" the veins were depofited by water in the moft
" perfect tranquillity, it is wonderful that we do
" not find thofe materials difpofed in horizon-
" tal layers acrofs the vein, inftead of being pa-
" rallel to its fides; and it feems very unac-
" countable, that the common ftrata, depofited
" as we are told while the water was in a ftate
" of great agitation, have fo rigoroufly obeyed
" the laws of hydroftatics, and acquired a paral-
" lelifm in the planes of their ftratification, which
" approaches fo often to geometrical precifion;
" while the materials of the veins, in circumftan-
" ces fo much more favourable for doing the
" fame, have done nearly the reverfe, and taken
" a pofition often at right angles to that which
" hydroftatical principles require. This is a pa-
" radox, which the Neptunian fyftem has creat-
" ed, and which therefore it is not very like-
" ly to refolve *."

This objection it is very eafy to remove; and
it is fomewhat fingular that Profeffor Playfair
fhould have ftated it in fuch ftrong terms, and
as even amounting nearly to a demonftration
againft the Neptunian theory, when the anfwer

* Illuftrations, &c. p. 251, 253.

M

is fo obvious and fatisfactory. The matter filling veins is perhaps always cryftallized, or has
the cryftalline ftructure, fo as to fhew that it has
been by this fpecies of confolidation that it was
formed. Now cryftallization is always promoted by a *nucleus* or fupport, and in any cavity, invariably takes place from the bottom and fides.
In the cafe of a vein therefore filled with a folution of different kinds of matter, the fubftance
moft difpofed to cryftallize would firft form an
incruftation on the fides and bottom, and afterwards, thofe which had lefs difpofition under the
circumftances of the cafe to cryftallize. According to thefe circumftances too, the fubftances
prefent might be varioufly mixed. Werner
ftates this very fact of the incruftation of the
materials of veins on their fides as a proof of his
theory, and particularly, that the incruftations on
one fide are always alike and difpofed in the
fame order, as on the other fide of the vein. It is
remarked alfo, as highly favourable to the fame
theory, that the different coatings are of a greater thicknefs at a depth in the vein, than they
are nearer to the furface.

If, on the other hand, the vein had been filled with different kinds of matter in fufion by
injection, it is evident, that thefe ought immediately to begin to arrange themfelves in the order of their fpecific gravities; the confolidation

of them from cooling could not commence immediately; and therefore, while still fluid, the heavier substance, the metal, for example, should fall towards the bottom, and the much lighter matrix rise to the top, a disposition which is never observed : And, when they began to concrete, it is equally obvious that the one least fusible should first become solid, while the more fusible remaining fluid, would be forced aside and collected apart; and thus there could be none of that intermixture, and those mutual impressions, which are generally observed in the materials of a vein; nor could the cooling of that part of the matter which was next to the sides of the vein, (a cause stated by Mr Playfair), have much effect in preventing this, or in disturbing the regularity of consolidation. This hypothesis, therefore, does not account for the appearances which veins actually exhibit.

It may also be observed, as an important fact in geology, that veins have occurred in which the depositions are horizontal. De Luc gives an example of this kind. The fact is utterly inconsistent with the Huttonian theory; while it proves clearly that the vein has been filled by deposition from above, the deposite being probably more mechanical than chemical, and therefore taking this form.

It is stated as an argument in favour of the

M 2

Huttonian account of the formation of veins,
that they " contain abundant marks of the
" moſt violent and repeated diſturbance," ſhifting of the ſtrata, and ſhifting or heaving of the
vein itſelf.

Theſe appearances, however, are likewiſe very
eaſily explained by the Wernerian theory : the
explanation has indeed been already ſtated, in
ſhowing the fallacy of the argument for the igneous origin of whin, from the derangements
in the ſtrata which accompany its veins. All
veins ariſe from fiſſures in the ſtrata or rocks in
which they are ſituated ; theſe fiſſures have a-
riſen from unequal ſinkings of theſe ſtrata, ſoon
after their formation, and have taken place ſucceſſively. It is obvious, therefore, that they
muſt often be accompanied with thoſe very
marks of diſturbance enumerated in the objection, and of courſe, that theſe do not prove
veins to have been formed by an irruption of
fluid matter from beneath.

Werner has ſtated the principal varieties of
diſlocation which attend veins, has ſhown that
they admit of the moſt ſatisfactory explanations
from his theory, and that the concluſions ſug-
geſted by the theory, lead to important prac-
tical applications in the art of mining. An order is obſerved even in theſe derangements,
which may be connected by a principle ; while,

were the Huttonian hypothefis true, there fhould be nothing but ruin and diforder.

" The fact of pieces of rock being found in-
" fulated in veins, is certainly favourable," it is faid, " to the notion of an injected and ponder-
" ous fluid having originally fuftained them." Admitting that fuch pieces reft in no part on the fides of the vein, the phenomenon is very eafily explained, from the fact, proved by vari- ous appearances, and admitted in the Huttonian theory, that veins have frequently been filled, not entirely at once, but fucceffively ; the fub- ftances next to the fides have been firft depofit- ed, and thofe in the middle of the vein have often been of more recent formation. It is ob- vious, that if, after a partial incruftation of the vein, by which its diameter would be diminifh- ed, fragments of rock were brought, by the cir- culation of the furrounding fluid, or detached by fome violent finking, or fubverfion of the rock itfelf, and introduced into the cavity, they might be fuftained by the fides, and the vein being afterwards filled up by new depofi- tions, the appearance of an infulated fragment would be produced. It is alfo poffible, that af- ter a vein has been filled, and completely con- folidated, a new fiffure might take place in it, preferving the fame direction ; and pieces of rock falling into it, from either of the above

M 3

caufes, might be cemented by new matter depofited from water; and in this manner Werner has explained the origin of cemented fragments of this kind which occur in fome veins.

A fact in the ftructure of veins incompatible
with the Huttonian hypothefis, and proving the
Wernerian theory, of their having been filled
from above, is that of petrifactions, marine fhells,
and even vegetable fubftances, being frequently
found in the fubftances filling the vein, a fact
eftablifhed on the authority of Werner, and
others, and even allowed by Dr. Hutton*, nor
can there be imagined any more decifive. As the
matter with which veins are filled is fuppofed,
by the Huttonian, to be thrown up in perfect
fufion, it cannot be conceived that fuch fubftances fhould be brought up from the central
parts of the globe in this ftream of melted
matter, or that if they had fallen into it, they
fhould have fuffered no change; while the explanation of their origin is obvious and natural,
when we fuppofe the matter in which they are
inclofed to have been depofited from the fea.

An argument of a fimilar kind may be derived from the fact, that veins are fometimes
filled with fubftances which obvioufly could
never have been in fufion. Thus Werner men-

* Theory of the Earth, vol. i. p. 396.

tions a vein at Riegelsdorf, in Heffe, the materials of which are nothing but fand and rounded
ftones. And Schreiber cites a vein in a mountain near Allemont, filled with an argillaceous
earth and rounded fragments of gneifs, and
which intercepts the metallic veins *. Nothing
can more certainly prove, that veins are filled
from above, and not formed by irruptions from
beneath.

The alteration and decompofition of the rock
at the fides of the vein, have been already noticed, as being explained more fatisfactorily by
the Neptunian than the Huttonian theory. The
decompofition of granite, hornblende, and
gneifs, into a clay or foft earth, for feveral inches
by the fide of the vein, cannot be explained,
but from the percolation and chemical action
of a fluid which has filled the vein.

In the connection of the contents of veins
with certain ftrata, an order is obferved, inexplicable in the Huttonian theory, but fatisfactorily explained in the Neptunian. This has
been traced by Werner. Tin is never found
but in primary ftrata, principally in granite.
Molybdena and Tungften are found in the fame
fituations, and of courfe have been formed at
the fame period. Uranium and Bifmuth, though

---

* Journal des Mines, No. xviii. p. 71.

perhaps of a formation rather lefs ancient, appear never to be found in ftratified mountains. Gold and Silver are fometimes found in the latter, though rarely. M cury, the grey ore of Antimony, and Manganefe, are difcovered both in primitive and fecondary mountains. Copper, Lead, Zinc, and efpecially Iron, belong to all the ages of the world. Cobalt and Nickel, are generally of recent formation. There is the fame difference to be obferved in the fubftances which accompany the metals. Felfpar, fhorl, the topaz, and the beryl, are confidered by Werner as the moft ancient. Quartz belongs to all periods. Among the calcareous fubftances, the moft ancient are fluor fpar, and apatite. Trap is of much more modern formation, and gypfum one of the moft recent *.

Now it muft be apparent, that the Huttonian hypothefis of the origin of veins can furnifh no principle by which any order of this kind can be explained. Thefe materials are fuppofed to be thrown from the central regions, in which they had exifted in a ftate of fufion, and their irruption muft have been merely accidental. No caufe can poffibly be imagined why certain metals fhould have been thrown up only in thofe rents which were fituated in the primary rocks ; others

* Journal des Mines, No. xviii. p. 90.

in the fecondary, and a third clafs in ftrata of
both kinds : but in the Neptunian fyftem a principle can be difcovered by which this may have
been regulated. The depofites from the chaotic
fluid, it is proved, were fucceffive, and it muft,
of courfe, have happened that fome kinds of
matter would be formed and depofited at one
period, others at another, from the play of affinities exerted, and the force with which they were
held in folution. In other words, the fame caufe
which determined granite to be firft formed, may
have determined the formation or depofition of
tin, molybdena, and tungften, at the fame time ;
and this principle may be extended to the production of all the others. It is a merit, in a theory,
of no trivial importance, that it fhould thus be
able to connect, by one principle, facts of the
firft importance, but apparently fo difficult to be
explained.

Laftly, there are veins to which the Huttonian
hypothefis cannot poffibly apply—thofe which
are included in rocks, and fhut in on all fides.
Thefe cannot be fuppofed to have been filled by
injection, as the termination of them in the rock
is obvious. Profeffor Playfair is obliged to relinquifh the general theory, and to fuppofe that when
" thefe veins are found in ftratified rocks, fuch as
" have not themfelves been melted, we muft con-
" ceive them to be compofed of materials more

" fufible than the furrounding rock, fo that they
" have been brought into fufion by a degree of
" heat which the reft of the rock was able to re-
" fift, and on cooling have affumed a fparry
" ftructure. When they are found in rocks of
" which the whole has been fluid, they muft be
" confidered as component parts of that mafs,
" which by an elective attraction have united
" with one another, and feparated themfelves
" from the fubftances to which they had lefs af-
" finity *."

The firft of thefe explanations, that which regards thefe infulated veins in ftratified rocks,
may probably afford a proof of the falfity of the
general hypothefis on this fubject. The matter
cryftallized in thefe veins is generally either
quartz, or carbonate of lime. Now, there is no
fubftance exifting in the form of a ftratified rock
lefs fufible than either of thefe foffils : they confequently could not have been " brought into
" fufion by a degree of heat which the reft of
" the rock was able to refift;" and therefore the
phenomenon of a vein of fuch fubftances in ftratified rocks cannot be accounted for on the principles of the Huttonian theory.

It is found alfo that thefe infulated veins fometimes contain metallic ores ; and, indeed, many

* Illuftrations, &c. p. 259.

metallic veins have been completely worked
out; yet Dr. Hutton is pleafed to tell us that
thefe can be derived only from the bowels
of the earth. " Look into the fources of our mi-
" neral treafures : afk the miner from whence
" has come the metal into his vein. Not from
" the earth, or air above; not from the ftrata
" which the vein traverfes; thefe do not contain
" one atom of the minerals now confidered.—
" There is but one place from whence thefe mi-
" nerals may have come, this is the bowels of
" the earth, the place of power and expanfion,
" the place from whence muft have proceeded
" that intenfe heat by which loofe materials have
" been confolidated into rocks, as well as that
" enormous force by which the regular ftrata
" have been broken and difplaced *."

    " The above (adds Profeffor Playfair) is a very
" juft and natural reflection; but if, inftead of
" interrogating the miner, we confult the Neptu-
" nift, we will receive a very different reply. As
" this philofopher never embarraffes himfelf
" about preferving an uniformity in the courfe
" of nature, he will tell us, that though it may be
" time that neither the air, the upper part of the
" earth's furface, nor even the fea, contain at
" prefent any thing like the materials of the

*Theory of the Earth, vol. i. p. 130.

" veins, yet the time was when thefe materials
" were all mingled together in the chaotic mafs,
" and conftituted one vaft fluid, encompaffing
" the earth; from which fluid it was that the
" minerals were precipitated, and depofited in
" the clefts and fiffures of the ftrata *."

After this declamation, it is unfortunate that
the Huttonian fhould be compelled to admit
that there are examples of veins filled with metallic matter, cut off from that fource from
which only, according to Dr Hutton's explanation, they could have been filled, and of which
no probable explanation can be given but that
by the Neptunian theory. It is even probable
that all veins are of this infulated kind, their
limits not being always difcovered, from their
not being explored to fufficient depth ; for it is,
prima facie, an extreme improbability that
rents fhould pafs through innumerable ftrata
even to the central parts of the globe, and that
thefe fhould be filled by injection to the very
furface, and through crevices often only a few
lines in diameter for a confiderable length.

An appearance not uncommon in veins is that
of their becoming narrower as they defcend, of-
ten leffening to $\frac{1}{4}$ of an inch or lefs in thicknefs,
and either remaining fo or again becoming wider,

* Illuftrations, &c. p. 248.

while above, they are perhaps fome feet in wide-
nefs. It is obvioufly impoffible that the matter
filling this upper part fhould have been thrown
contrary to its own gravity, by injection, through
fuch a paffage, from the central parts of the
globe.

PART IV.

Of the Support which the Huttonian and Neptunian Theories derive from the Appearances and Properties of Individual Foſſils.

This part of the inveſtigation of theſe theories, from induction, is that perhaps from which we may derive the moſt concluſive evidence. The poſitions and connections of the ſtrata, can, in many caſes, be only imperfectly obſerved, and they are alſo often ſuch as might have ariſen from various cauſes. But it is reaſonable to preſume, that in individual minerals properties will be found which ſhall afford undoubted proof whether they have been formed by water or by fire.

In conſidering the examples of this claſs, brought forward in proof of the Huttonian theory, I ſhall take them in the order in which they are ſtated by Profeſſor Playfair. The firſt are thoſe belonging to the SILICEOUS GENUS.

" Foſſil-wood, penetrated by ſiliceous matter,

" is a fubftance well known to mineralogifts;
" it is found in great abundance in various fi-
" tuations, and frequently in the heart of great
" bodies of rock. On examination, the filice-
" ous matter is often obferved to have pene-
" trated the wood very unequally, fo that the
" vegetable ftructure remains in fome places
" entire ; and in other places is loft in a homo-
" geneous mafs of agate or jafper. Where
" this happens, it may be remarked, that the
" line which feparates thefe two parts is quite
" fharp and diftinct, altogether different from
" what muft have taken place, had the flinty
" matter been introduced into the body of the
" wood, by any fluid in which it was diffolved,
" as it would then have pervaded the whole,
" if not uniformly, yet with a regular gradation.
" In thofe fpecimens of foffil wood that are
" partly penetrated by agate, and partly not
" penetrated at all, the fame fharpnefs of ter-
" mination may be remarked, and is an ap-
" pearance highly characteriftic of the fluidity
" produced by fufion *."

From the appearance of foffil filiceous wood,
the unbiaffed obferver would be much more
ready to infer, that it had been petrified by the
operation of water, fince it is fcarcely poffible
to believe that it could be fubjected to the action of fire, and at the fame time have preferv-

* Illuftrations, &c. p. 25.

ed completely the ligneous texture. It may be
conceived, that if a piece of wood be immerſed
in water, which holds diſſolved a portion of
ſilex, as many waters do, the earth may gradually be depoſited in its pores. At the ſame time,
from the ſlow putrefaction, or decompoſition,
which the wood in ſuch a ſituation muſt ſuffer,
its principles may be diſſipated in new products;
and if theſe two operations, the depoſition of
the ſiliceous matter, and the decompoſition of
the wood, bear a certain proportion to each
other, the earth will be depoſited in the vacuities left by the vegetable matter; and thus an
arrangement will be preſerved ſimilar to that of
the ligneous fibre. Hence might ariſe the peculiar character of foſſil ſiliceous wood;—its
being entirely deſtitute of vegetabe matter,
while it preſerves the texture of the wood, often
ſo perfectly, that the particular ſpecies can be
diſcovered.

But how ſhall we account for theſe effects,
if we are to ſuppoſe the wood to have been
ſubjected to the action of melted ſiliceous matter? How could this matter have penetrated
the ſubſtance of the wood? Still more, how
could it penetrate it, ſo as to preſerve the ligneous texture, and even the delicate reticulated
ſtructure? How could the ligneous matter have
been removed, while the ſiliceous was depoſited

in its place? The operation of fufion, by which
foffils are formed, is fuppofed, in he Huttonian
theory, to take place under a vaft compreffion;
this would prevent the volatilization or decompofition of any part of the wood, or if it did
not, according to the principles of the fyftem itfelf, the wood fhould have been charred
or converted into coal. In fhort, it feems impoffible to give, it need not be faid a fatisfactory
explanation, but any explanation whatever, of
the properties of filiceous wood by this hypothefis; on the contrary, were it true, had wood
been expofed to the action of melted flint, it
muft either have been decompofed and charred
by it, or muft have encrufted it, forming around
it a homogeneous indeftructible mafs.

The particular appearance in fome fpecimens
of the wood, which has been ftated as favourable to the Huttonian hypothefis, that of the
line of feparation between the part that is petrified and the part that remains unchanged being fharp and diftinct, inftead of affording any
prefumption of the filiceous matter being introduced in a ftate of fufion, is not even explained
by that fuppofition, for we do not perceive why
the fufed matter fhould have terminated abruptly, fo as to prefent this diftinct line. Were the
Huttonian required to explain this circumftance,
he would, in truth, be puzzled for an anfwer,

N

and, inftead of affording any fupport to his opinion, it is an additional difficulty, which he is
unable to remove. The appearance feems to
be owing to the procefs in the humid way going
on very flowly, probably from the denfity of the
wood, and to its having been ftopt in thofe fpecimens before it was complete. It is conceivable that the external part of the wood may
have been fo completely impregnated with the
filiceous matter as to prevent the infiltration of
the water to the internal parts, and of courfe
the procefs of petrifaction would ceafe, and this,
if the wood has been very denfe, it is poffible
may have proceeded only a fhort way, and have
been abruptly terminated.

There is another fact, with refpect to filiceous wood, which gives indication of its watery
origin, while it is inexplicable on the Huttonian
hypothefis; it is, that of fhells often adhering to,
and even indented in the wood, deprived of
their calcareous matter, and thus forming filiceous petrifactions. The prefence of thefe proves
that the wood had been immerfed in water; and
the entire converfion of their fubftance into filiceous matter cannot be accounted for on the
fuppofition of melted filex being applied to
them; for by what power was this filex to expel the calcareous carbonate, of which they
principally confift. On the oppofite opinion

this admits of explanation, the carbonate of lime, being more foluble in water than the filiceous earth, would be gradually carried off, and the latter might be depofited in its place.

In favour of the petrifaction of filiceous wood being effected by the medium of water, we have the analogous cafe of the petrifaction of wood in this way by calcareous matter. There are many fprings, in which a portion of carbonate of lime is held in folution, which have the power of petrifying any vegetable fubftance thrown into them; and many examples of petrified moffes and other matters are to be found, which no one can fuppofe to be performed by fire.— Here the ultimate effect is the fame. The only difference is, that in the one cafe the petrifaction is filiceous, in the other calcareous. Does not the prefumption follow, that the procefs by which it has been effected is fimilar; that in the one, calcareous matter diffolved in water; in the other, filiceous matter in folution has been applied to the petrified fubftance?

Laftly, we have demonftration that filiceous petrified wood is formed in the humid way. Mr. Kirwan relates a decifive proof of this kind: one of the timbers fupporting Trajan's bridge over the Danube being taken up, and examined, was found to have been converted into agate to the depth of half an inch, while the inner parts were more flightly petrified *.

With fuch weight of evidence, the conclufion cannot be refifted, that filiceous wood has been petrified by the medium of water; and this firft proof of the Huttonian theory, from the properties of minerals, ferves to eftablifh the oppofite fyftem.

This argument with refpect to the filiceous petrifaction of wood, is of more importance than at firft view may appear. It is not merely in itfelf an example which may be brought in fupport of the one theory or the other, neither is its importance confined to its proving the folubility of filiceous earth in water, and the poffibility of the moft perfect confolidation being affected by its depofition; but it eftablifhes a fimilar formation with regard to other foffils, and furnifhes a proof capable of being carried to a confiderable extent. It is obferved in fpecimens of wood thus changed, that where there are rents or vacuities in the wood, the filiceous matter depofited in thefe, has always affumed the figure and ftructure of agate. It has the concentric coats of that foffil, its hardnefs, frequently its various fhades of colour, and in fhort, all its properties. If, there-

Geological Effays, p. 110.

fore, the petrified wood is proved to be formed
in the humid way, it follows, that agates may
be formed in the fame mode ; and this fact a-
gain may be pufhed ftill farther, for agates are
almoft always found inclofed in other rocks, as,
for example, in trap, and inclofed in fuch a
manner as to render it undoubted that the rock
and the inclofed agate muft have had the fame
origin. So far, therefore, may the application
of this argument be carried, from the intimate
connection of thefe foffils. Dr. Hutton might
perceive this connection, and the obligation it
laid him under of afcribing the formation of filiceous petrified wood to fufion, fince, if he admitted it to have been formed in the humid
way, he muft have been forced to admit, that
agates and the rocks in which they are inclofed,
might have had the fame origin. It was, perhaps, a proof of polemical fkill to affume that
as an argument which he might have otherwife
been obliged to obviate as an objection ; and
this might lead him to maintain, that filiceous
petrified wood was formed from the introduction
of fufed filiceous earth, though its appearance is,
prima facie, inconfiftent with that opinion, and
though it is fully refuted by facts.

" The round nodules of flint that are found
" in chalk, quite infulated and feparate from
" one another, afford," it is faid, " an argument

N 3

" of the fame kind; fince the flinty matter, if
" it had been carried into the chalk by any fol-
" vent, muft have been depofited with a certain
" degree of uniformity, and would not now ap-
" pear collected into feparate maffes, without
" any trace of its exiftence in the intermediate
" parts. On the other hand, if we conceive
" the melted flint to have been forcibly injected
" among the chalk, and to have penetrated it,
" fomewhat as mercury may, by preffure, be
" made to penetrate through the pores of
" wood, it might, on cooling, exhibit the fame
" appearances that the chalk-beds of England
" do actually prefent us with *."

This theory of the formation of flint is nearly inconceivable, and is inconfiftent with the
appearances of that foffil. The kind of injection, by which it is fuppofed to be introduced
into the chalk, is altogether myfterious; for we
perceive not, how, without any fenfible open-
ings, the flint is to be conveyed into the chalk,
or if it were forced in by fome peculiar exertion of preffure, how the particles are to be collected, fo as to form nodules of confiderable
fize. The arrangement of thefe nodules is alfo incompatible with any notion of this kind.
They are not irregularly interfperfed in the

* Illuftrations, &c. p. 25.

chalk, as they neceffarily muſt have been, if
introduced by any ſpecies of injection, but are
found arranged with the greateſt uniformity, in
horizontal beds or layers, which are equidiſtant
from each other.

That the formation of theſe nodules is owing
to the agency of water, is proved by the impreſſions of ſhells on their internal ſurface, or even
entire ſhells inhering in the ſubſtance of the flint,
completely petrified, and having their calcareous
matter entirely abſtracted. It is obvious, that
this petrifaction and expulſion of the calcareous
matter, while the ſtructure and diviſions of the
ſhell often remain unaltered, could never be
produced by injection of fuſed ſiliceous matter, while they may, in common with other petrifactions, be eaſily explained from the agency
of water. There is a particular ſpecimen of
this kind, one of the moſt common, which, perhaps, places this in the cleareſt light,—that of the
ſhell of the Echinus, which is frequently found
filled with flint, while its texture remains perfectly unaltered; and its calcareous matter is,
at the ſame time, ſo completely removed, that
it does not effervefce on the application of an
acid. This ſhell is, in its natural ſtate, ſo extremely tender, that the leaſt preſſure cruſhes it
to pieces; and it has only a ſmall aperture lead-
ing to its cavity. Can it be ſuppoſed, that it

N 4

has been filled with fused flint, by any species
of injection? What would be the circumstances
which must have been present for such an effect
taking place? That the shell should have been
placed in such a manner that its aperture should
lie in the direction of the stream of injected
fluid, and that it should have such firmness, as
to suffer no alteration from being completely
filled with this dense fluid matter; and were
even the supposition of the existence of these
circumstances admitted, the expulsion of the
calcareous matter of the shell would remain inexplicable. This, and other petrifactions, evidently show, that the particles of flint have been
brought together slowly, and without violence,
by the agency of water. Mr. Kirwan attributes their consolidation, to the infiltration of
that fluid through the strata of chalk *; and
their formation into nodules, may be ascribed,
according to the Wernerian hypothesis, to the
particles being collected in cavities formed in
the chalk by the extrication of air.

" The siliceous pudding-stone," it is said, " is
" an instance closely connected with the two
" last; in it we find both the pebbles, and the
" cement which unites them, consisting of flint
" equally hard and consolidated; and this cir-

* Geological Essays, p. 237.

" cumftance, for which it is impoffible to ac-
" count by infiltration, or the infinuation of an
" aqueous folvent, is perfectly confiftent with
" the fuppofition, that a ftream of melted flint
" has been forcibly injected among a mafs of
" loofe gravel *."

There feems nothing improbable in the fuppofition that the pebbles in this foffil have been
agglutinated and confolidated by the depofition
of the cement from aqueous folution or fufpenfion. Suppofe thefe pebbles to have been
placed in a fituation fimilar to that of petrified
filiceous wood ; its pores have been completely
filled with filiceous matter, of the nature of a-
gate ; and, in like manner, the interftices between thefe loofe pebbles may have been filled
with this matter, and thus a confolidated hete-
rogeneous mafs might be formed.

Agates belong to the filiceous genus, and
are confidered by Dr. Hutton as affording, from
their ftructure, an argument in favour of his
theory. They are maffes, the internal parts of
which have a certain arrangement, which, it is
alleged, could not be produced by the infiltration of any fluid, but which muft have arifen
from the circumftances under which the agate
was formed.

* Illuftrations, &c. p. 26.

That the fluidity from which agates have been confolidated, has been that of fufion, it is faid, is evident, becaufe " the formation of the " concentric coats, of which the agate is ufual- " ly compofed, has evidently proceeded from " the circumference toward the centre, the ex- " teriot coats always impreffing the interior, " but never the reverfe. The fame thing alfo " follows from this other fact, that when there " is any vacuity within the agate, it is ufually " at the centre, and there too are found the re- " gular cryftals, when any fuch have been form- " ed. It therefore appears certain, that the pro- " grefs of confolidation has been from the cir- " cumference inwards, and that the outward " coats of the agate were the firft to acquire " folidity and hardnefs.

" Now, it muft be confidered that thefe " coats are highly confolidated; that they are " of very pure filiceous matter, and are utterly " impervious to every fubftance which we know " of, except light and heat. It is plain, there- " fore, that whatever at any time, during the " progrefs of confolidation, was contained with- " in the coats already formed, muft have re- " mained there as long as the agate was entire, " without the leaft poffibility of efcape. But " nothing is found within the coats of the a- " gate fave its own fubftance; therefore no ex-

" traneous fubftance, that is to fay no folvent,
" was ever included within them. The fluidity
" of the agate was therefore fimple, and unaf-
" fifted by any menftruum.

" In this argument, nothing appears to me
" wanting, that is neceffary to the perfection of
" a phyfical, I had almoft faid of a mathemati-
" cal, demonftration. It feems, indeed, to be
" impoffible that the igneous origin of foffils
" could be recorded in plainer language, than
" by the phenomenon which has juft been de-
" fcribed *."

Notwithftanding the ftrong language in which
thefe affertions are expreffed, it may be demonftrated, that agates could not have been
formed by fufion, but muft have originated
from the operation of water. Their ftructure is fo peculiar that there may be fome difficulty in pointing out clearly the precife mode
of their formation, but the following explanation by Werner, and which has alfo been fuggefted by Dolomieu and Kirwan, will be found
to accord better with the appearances of the
foffils of this family, then the Huttonian hypothefis.

During the confolidation of the ftrata, it is
fuppofed that extrications of air muft have taken

* Illuftrations, &c. p. 78, 79.

place, which in a foft mafs would form vacuities of a fpheroidal form; after the confolidation had been completed, thefe vacuities are
fuppofed to have been filled with a fluid holding the matter of chalcedony, jafper, and other
fubftances of which agates confift, in folution.
Thefe would be depofited fucceffively from the
various affinities of the fubftances diffolved to
the folvent; and in this manner the fucceffive
coats would be formed, the exterior coat deter-
mining in a great meafure the figure of the interior depofited upon it. Thefe coats, it has
been obferved, vary in the purity or homogeneous nature of their fubftance, the outer is compofed of the coarfeft or moft heterogeneous fubftance, as of jafper or carnelion; as it approaches to the centre, it generally becomes purer, till
at length from the folution thus purified as it
were, by thefe depofitions, cryftals of quartz and
amethyft fhoot. When the folid matter had
been completely feparated from the fluid, the
latter would efcape by percolation (for the hardeft foffils of this clafs, contrary to the affertion
of Dr. Hutton, are capable of allowing water to
pafs through them): if the orifice by which the
folution entered was clofed up by the depofition, a hollow agate would be formed; if it
were not, fucceffive portions of the folution

might find accefs, and the whole cavity be completely filled.

From this hypothefis, the general appearances of agates can be explained; and there is one
variety, that of the hollow agate, which can
fcarcely be explained by any other, or which at
leaft can receive no explanation from the Huttonian hypothefis. The cavities in thefe agates
are fuppofed, according to the latter hypothefis,
to arife from the contraction of the mafs in its
confolidation and cooling. But there are many
of them in which the cavity is fo difproportion-
ed to the folid cruft, that it could not poffibly
have arifen from this caufe. Thus they may be
found in which the cavity is four inches in
diameter, while the folid external coat is not
more than $\frac{1}{4}$ inch thick : it is obvious that no
fhrinking of this mafs could have produced a
cavity of this kind. The explanation again by
the Wernerian hypothefis, in which the production of the cavity is referred to the extrication of a portion of ærial fluid, and the fubfequent depofition of the matter of which the
cruft is formed, is perfectly fatisfactory ; at the
fame time it cannot be adapted to the Huttonian hypothefis, for according to that fyftem,
the fufion of the ftrata has taken place under
an immenfe compreffion, by which every extrication of air is prevented, and fuch a preffure

muft, even by the defender of that hypothefis, be fuppofed prefent in the formation of thefe hollow agates, becaufe they frequently contain carbonate of lime, which, but for this circum-ftance, muft have been decompofed.

There is another appearance in certain agates which Dr. Hutton adduces as an argument in fupport of his theory, and which affords a proof of its falfity. In agates of chalcedony, he ob-ferves, calcareous fpar is often inclofed, and thefe are found mutually impreffing and im-preffed by each other; " the angles and planes " of the fpar are indented into the chalce- " dony, and the fpherical fegments of the chal- " cedony are imprinted on the planes of the " fpar. Thefe appearances are confiftent with " no notion of confolidation that does not in- " volve in it the fimultaneous concretion of the " whole mafs; and fuch concretion cannot a- " rife from precipitation from a folvent, but " only from the congelation of a melted " body *."

Precifely the oppofite conclufion may be drawn, with much more juftice,—that this fimul-taneous confolidation could not take place from the fluidity produced by fufion; for fubftances of different natures have always different degrees

* Illuftrations, &c. p. 247.

of fufibility, and, having this difference, muft
become folid at different temperatures. Either
the chalcedony muft have been more fufible
than the calcareous fpar, or the fpar than the
chalcedony ; and, which ever be the cafe, the
one that was leaft fufible muft have concreted
firft, and might take its peculiar form, but
they never could concrete fimultaneoufly fo as
to imprefs and be impreffed by each other ; and
fince Dr Hutton not only allows but contends
for this mutual impreffion, he eftablifhes a fact
which overturns his own hypothefis of the
formation of thefe foffils.

It is indeed fingular that this cafe of fimultaneous confolidation or cryftallization fhould
have ever been employed as a proof that it had
taken place from fufion, as in various other
cafes befide that of agates it is, for it implies a
fuppofition which it is eafy to demonftrate is
falfe. If two or more different fubftances be
in fufion, it is evident that they will become folid, according to their fufibilities ; the leaft fufible, requiring the higheft temperature to preferve
it folid, will, on a reduction of temperature, firft
pafs to the folid ftate, and will be fucceeded by
the others, which are more fufible. It cannot
be fuppofed that two fubftances fhould, from a
ftate of fufion, become folid precifely at the fame
time, fo as to imprefs each other, unlefs they

are of precifely the fame degree of fufibility. Now there are not, perhaps, two fubftances in nature with refpect to which this is the cafe. At leaft, if it happened with any two, it might be confidered, both as a fingular coincidence, and ftill more as the refult of a very extraordinary combination of circumftances, that of all bodies the two which thus happened to agree in fufibility fhould have been brought together in fufion. But it may fafely be affirmed, that not another example of it would be met with, far lefs that it fhould take place with refpect to many fubftances; and if it fhould actually be found in nature that there were a variety of groupes or aggregates of foffils which mutually impreffed each other, this, inftead of being regarded as a proof that thefe fubftances had confolidated from fimple fufion, would furnifh the cleareft demonftration that the fluidity from which they had become folid muft have been of a different kind.

It may be faid, perhaps, that it is not clear how from folution two different fubftances fhould confolidate or cryftallize at the fame time. Let it be granted, that it is not obvious how this fhould happen; yet ftill we have no demonftration, as in the former cafe, that this could not be the cafe; and therefore, as it is the only fuppofition left to us, it ought to be admitted. But

we need not reft fatisfied with this argument. It
is poffible to conceive how different fubftances in
folution in one fluid may be brought either to
cryftallize or confolidate together, or to give the
appearance of fimultaneous confolidation. Let
it be imagined, that by fome alteration of circumftances in the folution the cryftallization of
one of the fubftances was occafioned : it is conceivable that the feparation of this fubftance
might fo alter the exifting attractions, by which
one or other of the remaining fubftances in the
folution were held diffolved, that it alfo might
inftantly begin to feparate, and thus its confolidation would be fimultaneous, or very nearly fo, with that of the other, and they might
imprefs each other : Or, as cryftallization is promoted by a *nucleus*, and ftill more by the prefence of a body already cryftallized, it is conceivable, that if in a faturated folution of different fubftances, one of them fhould by an alteration of circumftances, be brought to cryftallize,
the cryftallization of the others might inftantly commence, as we fee take place when we
drop a folid into a faturated folution of a falt;
and thus the moft intimate admixture and mutual penetration of cryftals might take place.
Nay, it is no improbable fuppofition, that by
the fudden admiffion of air to fuch a faturated
fluid, or even by the fudden evolution of any

O

gas from it, the inftantaneous cryftallization of more than one of the fubftances diffolved in it might be produced.

It is thus evident, that various fuppofitions fufficiently probable can be made, by which the fimultaneous confolidation of different fub-ftances from a ftate of folution may be account-ed for. It cannot, therefore, be faid, that this is a difficulty which cannot be explained, and which preffes againft the theory. On this point the Neptunian has every advantage over the Huttonian hypothefis. It can be demonftrated, that according to the principles of the latter no fuch thing as fimultaneous confolidation could take place. No demonftration of this kind can be brought againft the other, which at once, therefore, gives it the fuperiority. But it can claim a ftill higher advantage, fince it is even eafy to fhew, in perfect conformity with its prin-ciples, how this fingular operation of fimultane-ous confolidation might take place. Every inftance of it, therefore, fuch as that of chalce-dony and calcareous fpar in agates, and others to be afterwards noticed, are fo many proofs of the formation of foffils by folution.

" The common grit, or fandftone, though it
" certainly gives no indication of having pof-
" feffed fluidity, is ftrongly expreffive of the ef-
" fects of heat. It is fo, efpecially in thofe in-

" ftances where the particles of quartzy fand,
" of which it is compofed, are firmly and clofe-
" ly united, without the help of any cementing
" fubftance whatfoever *." The Neptunian
would fay, that this is mere affertion and no
argument whatever, as the confolidation of this
ftone, the denfity or compactnefs of which is
not great, might be produced by the mutual
attraction exerted between its finer particles;
or, as all fandftone contains a portion of argil
or lime, thefe might alfo ferve, in part at leaft,
as the connecting medium.

If we fhould, *a priori*, expect, from an examination of the ftrata of the earth, any indication of their origin, or of the nature of the a-
gents to whofe operation they had been fubjected, it is certain that none lefs ambiguous
could be obtained than that afforded by the
animal and vegetable impreffions and remains
which they contain. Thefe at once afford the
double proof, that they have been depofited
from water, and that they have not been fubjected to the action of fire.

This proof we fhall find in all the ftrata,
more fparingly perhaps in thofe of the Siliceous
Genus than in the others, but ftill in thofe fo
abundantly as to furnifh the moft conclufive

* Illuftrations, &c. p. 259.

O 2

evidence. Befides the petrifactions in flint, which have been already noticed, they occur in petrofilex, and filiceous fandftone. It is obvious, that the intenfe heat which, according to the Huttonian theory, is applied to thefe foffils for their confolidation, ought to have foftened or fufed the fubftance of thefe petrifactions. Granite and whin, in that theory, are fuppofed to be formed nearly from the fame materials as thefe ftrata,—from fubftances depofited at the bottom of the fea; and the reafon given why they contain no petrifactions, or no remains of marine animals, is, that by the heat by which the whin or granite has been fufed, they have been deftroyed. But the heat neceffary even to foften petrofilex, or filiceous fandftone, is much greater than that requifite to fufe whin; and therefore, *a fortiori*, thefe foffils ought not, more than whin, to contain organic remains. This follows ftill more ftrongly from the fact that the filiceous and calcareous earths act mutually as fluxes, fo that if a fhell were involved in fandftone, foftened by heat, this alone fhould caufe them to combine.

Among the foffils belonging to the CALCAREOUS GENUS, are ftated, as proving the action of fubterranean heat, " the calcareous breccias, " compofed of fragments of marble or lime- " ftone, and not only adapted to each other's

" fhape, but indented into one another, in a
" manner not a little refembling the futures of
" the human cranium. From fuch inftances,
" it is impoffible not to infer the foftnefs of
" the calcareous fragments when they were con-
" folidated into one mafs. Now, this foftnefs
" could be induced only by heat ; for it muft be
" acknowledged that the action of any other
" folvent is quite inadequate to the foftening
" of large fragments of ftone, without diffolving
" them altogether *." This appearance of thefe
marble breccias can fcarcely be accounted for
on the fuppofition of their being foftened by
heat ; for in the joinings of the fragments the
fharp angles of many of them are often preferved at their infertions. In general, thefe
fragments are connected and interlaced, as it
were, by a common ground or cement, and not
indented into the fubftance of each other, and
this connecting cement may have been depofited from water around thefe fragments.
Where there is an appearance of indentation in
fome places, it might arife from the accidental
pofitions of the fragments, and the preffure
from the gravity of the mafs, by which the
fharp projections have been inferted into hollows or fractures of the others, and united by

* Illuftrations, &c. p. 28.

O 3

the common cement. Nay, the juftnefs of the
affertion in the objection may be denied ; for it
is poffible, that by the continued application of
a folvent to thefe fragments, they might partly
be diffolved, partly foftened. They might thus
be indented together by preffure, and a cement
at the fame time formed by which they would
be confolidated into one mafs.

" In many other inftances it appears certain,
" that the ftones of the calcareous genus have
" been reduced by heat into a ftate of fluidity
" much more perfect. Thus, the faline or finer
" kinds of marble, and many others that have a
" ftructure highly cryftallized, muft have been
" foftened to a degree little fhort of fufion, be-
" fore this cryftallization could take place. Even
" the petrifactions which abound fo much in
" limeftone tend to eftablifh the fame fact; for
" they poffefs a fparry ftructure, and muft have
" acquired that ftructure in their tranfition from
" a fluid to a folid ftate *."

In the firft of thefe inftances, there is merely
the gratuitous affumption that the cryftallization producing the fparry appearances of thefe
marbles, muft have been from fluidity induced
by fufion, and not from fluidity occafioned by
folution. The latter affords an example of a very

* Illuftrations, &c. p. 29.

fingular mode of reafoning. It is acknowledged, that the petrifactions in limeftone muft have been fufed, becaufe they could only acquire their fparry ftructure in their tranfition from a *fluid to a folid ftate*, and of courfe the figure and the ftructure of the fubftances petrified, which are generally fhells, muft have been loft. But how could this figure be refumed when the fufed fubftance returned to the folid ftate. If a fhell in a mafs of limeftone had been in fufion, it is obvioufly impoffible that in becoming folid it could again affume its precife arrangement and figure.

The proof from the prefence of thefe petrifactions in calcareous ftrata, in favour of the Neptunian theory, is conclufive. Thefe ftrata are fuppofed to be formed in the bed of the fea, and probably chiefly from the decay of marine animals. Their fhells confift principally of carbonate of lime ; the animal matter which is mixed with this is evolved by putrefaction during their flow confolidation, and its place is fupplied by a portion of the carbonate held in folution by the furrounding water : from a fimilar depofition, mixed with fmaller portions of argillaceous earth, oxyd of iron and other fubftances, their confolidation is completed, and the more flow depofition of pure carbonate of lime, gives rife to the formation of the fpar and cryftals which thefe ftrata always contain.

O 4

In the Huttonian theory, the prefervation of thefe animal remains is a problem which is not folved. It is maintained, that confolidation cannot be effected but by fufion, partial or complete; and of courfe it is maintained, that thefe ftrata of limeftone and marble have been melted, or at leaft foftened by heat. It remains, then, to be explained, how the fhells of the marine animals, which thefe ftrata contain, have efcaped the action of this heat; how their figure and texture have not been deftroyed? Their compofition is the fame as that of the remaining matter of the limeftone or marble; and indeed thefe ftrata are fuppofed, by Dr. Hutton, to originate from the remains of fea animals. If thefe, therefore, were in a fufed or foftened ftate, ought not the included fhells to have fuffered precifely the fame change? The fufion of thefe ftrata muft, according to the Huttonian, have been nearly complete, for they have the fparry ftructure, and contain large veins and fpots of perfect fpar. It is impoffible to conceive, how fuch a degree of fufion fhould have taken place and thefe fhells not have been at leaft fo far foftened, as to have loft fomewhat of their figure, and of the fharpnefs and diftinctnefs of their angles and lines. Nay, the cavity of the fhell is often filled with regular cryftals of carbonate of lime, which, if formed by heat, muft have concreted from a state of perfect fusion: but this fused matter could not have remained in contact with the shell, without melting it completely or partially; and, therefore, such appearances, which, so far from being rare, are extremely common, are incompatible with such a suppofition.

The conclufion becomes still more evident, when we confider the extreme delicacy often obferved in thefe impreffions. The calcareous stone of Monte Bolca is known to every mineralogist, for the large and perfect impreffions of different kinds of fifh found in it. Thefe are numerous, the entire figure of the fifh is impreffed upon it, and with fuch accuracy, that the characters of a number of fpecies have been eftablifhed. It is abfolutely incredible that an impreffion of this kind could remain on a stone fubjected to an intenfe heat, that the matter of the fifh fhould not have fuffered the fmalleft decompofition, or even mechanical derangement; and the bare infpection of thefe fpecimens will be fufficient to convince the unprejudiced obferver, that thefe ftrata could not have owed their confolidation to fire: but there is no difficulty in conceiving, that water, by that flow agency above explained, might form fuch petrifactions.

Another fact refpecting thefe foffils, which
cannot be explained in the Huttonian hypothefis, has been ftated by Mr. Kirwan,—the abfence of phofphoric acid in the finer kinds of
marble and limeftone. The fhells of marine
animals contain a portion of phofphat of lime
in their compofition, as has been eftablifhed
by the experiments of Mr. Hatchet; this fubftance, Mr. Kirwan juftly remarks, is indeftructible
by heat; and, therefore, had thefe marbles been
formed from the remains of marine animals, they
ought to contain the full proportion of phofphat of lime, which thefe contain. The reply
which Mr. Playfair gives to this, is extremely
obfcure. To give the argument force, he obferves, it would be neceffary to prove, that phofphoric acid exifts in thefe limeftones which evidently confift of fhells in a mineralized ftate:
if thefe are found without phofphoric acid, it is
evident that the preceeding argument fails entirely. But this proof is not neceffary. The
marble in which no phofphoric acid has been
found, is fuppofed, by the Huttonian geologift,
to be formed from remains of marine animals
equally with any other; and the circumftance
of other marbles or limeftone, containing or not
containing phofphat of lime, is of no confequence. If it be found, that other limeftones

do not contain phofphat of lime, the argument
will indeed become more general; but, however
this may be, the objection, from the particular
fact with regard to the marbles which have
been analyfed, remains the fame. To the Neptunian theory, the objection does not apply, becaufe the fmall quantity of phofphat of lime
prefent may be carried off by water.

It is, laftly, to be remarked, with regard to
the calcareous foffils, that the evidence for the
formation, of at leaft fome of them by water, is
fo unexceptionable, that it muft be admitted
even by the Huttonian. Thus, calcareous cryftals are often found in fituations where they
muft neceffarily have cryftallized from a
folution of carbonate of lime in water. They
are thus met with lining the cavities of fhells,
fuch, for example, as the fhell of the Cornua
Ammonis, and many others. Thefe fhells
themfelves confift of carbonate of lime ; and of
courfe they could not contain carbonate of lime
in perfect fufion, without fuffering an alteration. The ftate of fluidity, therefore, from
which thefe cryftals have concreted, has not
been that from fufion, but muft have been that
of folution in water. Another example, not lefs
equivocal, is that of calcareous ftalactites. Thefe
have not only in general the fparry texture, but
are often covered with cryftals, and their for-

mation in the humid way is abundantly obvious. The force of this conclusion is evident. It is proved, that calcareous cryftals, different in no refpect from others found in nature, are, in fome cafes at leaft, formed from folution; that they are ever formed by the fufion of carbonate of lime, is a mere hypothefis, fupported by no direct evidence;—till fuch evidence, therefore, be adduced, the Neptunian has a right to regard that which he brings forward as conclufive, and to confider all calcareous cryftals as being formed in the fame mode; and to this conclufion nothing but a fimple poffibility can be oppofed. Even that conjecture would be, *a priori*, improbable, fince it would be fingular, if this foffil fhould be capable of being formed both by fire and water, and fhould receive from each precifely the fame form of cryftallization. And, if it be rendered probable, that calcareous cryftals are formed by folution, the fame mode of formation muft be affigned to almoft every foffil; for there are very few with which thefe cryftals are not fo intimately affociated, that the fame origin muft neceffarily be given to them all.

Among the foffils of the ARGILLACEOUS GENUS, the variety of argillaceous iron ore, named the feptaria affords an argument on which Dr Hutton laid much ftrefs. This foffil is generally

found in fpheroidal nodules, which when broken exhibit a fingular ftructure. The ground
or bafe confifts of argillaceous iron ore, but
this is divided into a number of fepta by veins
of calcareous fpar, which fhoot from the centre
of the ftone to the circumference, but do not
reach it. This proves, it is faid, that the calcareous matter has not been introduced into the
ftone from without by infiltration. " The only
" other fuppofition that is left for explaining the
" fingular ftructure of this foffil is, that the
" whole mafs was originally fluid, and that in
" cooling the calcareous part feparated from the
" reft, and afterwards cryftallized *."

It might be granted, that the ftructure of this
ftone proves that it muft have been in a fluid
ftate, and that in hardening, the contraction of
the bafe, and the feparation of the calcareous
matter, had taken place at the fame time, without its following as a neceffary confequence that
this fluidity muft have arifen from fufion. It
may have been a depofite from water, and the
fhrinking and feparation may have taken place
during its confolidation; or it might be fuppofed, with perhaps equal probability, that the argillaceous iron ore only had been depofited; that
in confolidating it had fhrunk and fplit internal-

' Illuftrations, &c. p. 26.

ly; and as this is actually a porous stone, if it were afterwards exposed to water having carbonate of lime in solution, the water might enter by infiltration through its substance, and the cavities be filled, as far as they extended, with calcareous matter. That these septaria have had an aqueous and not an igneous origin, is proved by their sometimes containing impressions of organic substances, particularly shells.

In the argillaceous strata, petrifactions are extremely frequent, more particularly shells and impressions of vegetables. These are found in clay, argillaceous shistus, argillaceous sandstone, argillaceous ironstone, and various others. The argument with respect to them, is the same as with regard to the other strata in which they are found. The operation of fire ought to have altered or destroyed these remains and impressions, nor is any cause pointed out in the Huttonian theory by which they could be preserved.

This argument is conclusive, whether we consider the deficiency of explanation in the one theory, or the satisfactory solution afforded by the other. Let the example of marine shells preserved in sandstone be taken. If that sandstone has been formed by deposition from water, it might happen that the shells of animals existing in that water, might be involved in the de-

pofite ; and, in this cafe, no change would happen to them, except perhaps their being more
or lefs impregnated with particles of the matter
in which they were depofited. This is accordingly the ftate in which thefe fhells are found,
and the actual appearance exactly correfponds
with that which the theory would lead us to
expect. If, again, the ftratum of fandftone were
confolidated by heat, it is impoffible to conceive
how that heat fhould operate without changing
the figure or ftructure of the contained fhells.
Nay, in this example, another circumftance is
prefent, which ftill more forcibly proves, that
heat could not thus be applied without producing fome change. Argillaceous fandftone
confifts principally of filiceous and argillaceous
earth : now thefe ferve as a flux to calcareous
earth, and caufe its fufion at a temperature much
lower than that requifite to fufe it when pure.
Shells of marine animals have this earth for their
bafis ; and had heat been applied to them, fur-
rounded by argillaceous fandftone, fuch a combination muft have been effected.

Or, let the argument be confidered with refpect to the vegetable impreffions, which are
often found on fhale or fhiftus. Thefe are fo
delicate and perfect, that the genus, and even
the fpecies of the vegetable, can often be determined, from the prefervation of the moft deli-

cate foliage and flower. It may be aſked of
the Huttonian, if theſe ſtrata were conſolidated
by fire, as he ſuppoſes, how the vegetable matter could be entirely removed, while the impreſ-
ſion ſhould remain perfect and unimpaired?
The vegetable matter ought undoubtedly to
have ſuffered decompoſition; for, even if the
eſcape of its volatile parts were prevented by
the preſſure preſent, ſtill its principles muſt, under ſuch a heat, have entered into new combinations, by which their ſtructure muſt have
been altered, or, at all events, by the fuſion or
ſoftneſs of the ſtone on which the vegetable
matter lay, its figure muſt have been changed.

Among the BITUMINOUS SUBSTANCES, Coal is
brought forward as affording a proof of the igneous origin of foſſils. Inſtead of doing ſo, it
will be found, like the preceding examples, to
afford the ſtrongeſt objections to the Huttonian
hypotheſis.

From the nature of this ſubſtance, and the
appearances of its ſtrata, there can be little
doubt that it is principally of vegetable origin.
The moſt probable theory of its formation is,
that vegetable matter carried to the ſea, has,
by the direction of currents, been depoſited in
banks, and that during this ſubmerſion it has
ſuffered that ſlow kind of decompoſition by
which the greater part of its principles have

been evolved in new combinations, while its carbon, with a portion of hydrogen, have remained; and this, mixed with more or lefs earthy matter depofited at the fame time from the ocean, has in its foft ftate been confolidated by the force of aggregation, and has formed coal. The decompofition by which this has been effected, is probably analogous to that which we know animal matter, when immerfed in water, fuffers. Its hydrogen, azot, and oxygen, enter into various combinations, forming gafes which efcape; and its carbon retaining, by a chemical attraction, a portion of hydrogen in combination, remains, forming a fpecies of fat. Carbon is ftill more abundant in vegetable than it is in animal matter; and this conftitutes the principal difference between them. Vegetable matter, however, is liable to fimilar decompofitions; and under the circumftances pointed out, it is reafonable to believe that changes of a fimilar kind, modified as to the refult by the difference in the proportions of its principles, will take place: in other words, its refidue will be more carbonaceous, but ftill with a proportion of hydrogen, fo as to render it more or lefs bituminous. We accordingly find, that wood, by immerfion in water, becomes firft brown, and then black; and the ligneous fibre, by flow decompofition, is com pletely converted into a black mould, in which carbon predominates. It is eafily conceivable, that this procefs being carried on under different circumftances, may proceed with various degrees of rapidity, and to a greater or lefs extent. Hence will originate different varieties of coal, fome being much more carbonaceous than others, while their compofition is alfo varied by the different quantities of earth depofited during their formation.

It is not improbable that fome fpecies of coal may likewife have a different origin. It may be fuppofed, and indeed muft be, that carbon, in common with other fimple fubftances, exifted in the chaotic fluid; and this carbon, combined with a portion of oxygen, may have formed a variety of coal which was precipitated. This probably forms the coal, which is infufible and very little inflammable, the mineral carbon, or anthracite, and which is confidered by many mineralogifts as very remote from a vegetable origin. From its analyfis it does not appear to contain any bitumen, its principles, according to the experiments of Dolomieu and Panzenberg, being pure carbon, or at leaft oxyd of carbon, filex, argil, and oxyd of iron * : and it is obferved by Dolomieu, that this foffil, as well as graphite or plumbago, is found in the primitive mountains,

* Brochant, t. 2. p. 81.

generally in veins. The fuppofition offered refpecting the origin of this variety of coal, is alfo
confirmed by the fact, that carbon is found more
or lefs oxydated in the compofition of feveral
primitive foffils and rocks.

Dr. Hutton, however, confiders coal as a fubftance formed by the operation of fubterraneous
fire ; and fuppofes that there are feveral appearances connected with it, which prove it to have
had fuch an origin. It is fomewhat difficult to
give a clear ftatement of his theory on this fubject ; but it feems to be comprehended under
the following propofitions :

Firft, It is conceived, that an intenfe heat has
been applied to vegetable matter at the bottom
of the ocean, and that at the fame time no great
degree of preffure has been prefent. By this
operation the vegetable matter has been charred, or converted into one fpecies of coal, while
the more volatile inflammable matter of the
wood has been difengaged. *2dly*, This volatil-
ifed matter being of a bituminous nature, is fuppofed to be diffufed in the ocean, and to be
" employed in forming other ftrata, which were
" then depofited at the bottom of the water."
To this would be added, " all the fuliginous
" matter that is formed in burning bodies upon
" the furface of this earth, which is firft deliver-
" ed into the atmofphere, but ultimately muft

" be fettled at the bottom of the fea *." And, *laftly*, Another fupply of bituminous matter is derived from the vegetable fubftances diffolved or fufpended in the water of the rivers, and brought to the ocean. The bituminous matter from thefe fources is fuppofed to be precipitated either alone, or with a portion of fine earth, likewife fufpended in the water of the fea; and this precipitate, being afterwards confolidated by fubterraneous heat, forms ftrata of pure foffil coal.

It would be an irkfome, and a very unneceffary tafk, to enter on the formal refutation of thefe various fuppofitions. A few obfervations on the moft palpable deficiencies of the theory will be fufficient.

It may, in the firft place, be obferved, that there are no direct arguments adduced to eftablifh the igneous origin of coal. The theory is attempted to be eftablifhed principally by arguments drawn from its connection with other foffils, which it is to be proved have had an igneous origin. Thus, becaufe " the beds of coal " are difpofed in the fame manner, and are al- " ternated indifcriminately with thofe of all the " fecondary rocks," it is inferred that they muft have been formed by the fame operation,

* Theory of the Earth, vol. I. p. 577.

and that this has been fufion by heat, a conclufion founded on nothing peculiar to coal but on
the evidence refpecting the origin of thefe fecondary rocks. They are alfo faid to be " tra-
" verfed like the other ftrata by veins of all the
" metals, of fpar, of bafaltes, and of other fub-
" ftances *." But here likewife the argument is
not direct, but depends entirely on its being
proved that thefe veins have been formed by fufion. Laftly, the coal ftrata are faid to " contain
" pyrites in great abundance, a fubftance that
" is, perhaps, more than any other the decided
" progeny of fire *." But this, equally with
the preceding arguments, refts on a proof with
refpect to the origin of a different fubftance,—a
fubftance which it will immediately be fhewn
there is every reafon to conclude has been formed in the humid way.

From this ftatement, it is obvious that no direct argument from the properties or appearances
of coal, is adduced in proof of its igneous origin.

Not only, however, is it unfupported; feveral
of the fuppofitions it involves are highly improbable. Thus, it is imagined that a great part
of the coal ftrata are derived from bitumen produced either by the expofure of wood to fubterraneous heat, by the burning of vegetable mat-

* Illuftrations, &c. p. 33.

P 3

ter at the furface, or by the folution or fufpenfion of the oily and inflammable parts of vegetables in the water of rivers; this bitumen from
thefe fources being fuppofed to be diffufed in
the ocean, and precipitated fo as to form ftrata.
But how is this matter to be collected in one
place, and, if collected, how is its precipitation
to be effected. It is lighter than water, efpecially fea water: it muft therefore remain at the
furface, and no caufe can be affigned for its being carried to the bottom.

The peculiar modifications of preffure fuppofed requifite for the igneous formation of coal,
furnifh another argument againft the theory.
" It muft be confidered," fays Dr. Hutton, " that
" while immerfed in water, and under infuper-
" able compreffion, the vegetable, oily, and refin-
" ous fubftances, would appear to be unalterable
" by heat; and it is only in proportion as cer-
" tain chemical feparations take place that thefe
" inflammable bodies are changed in their fub-
" ftance by the application of heat. Now, the
" moft general change of this kind is in confe-
" quence of evaporation, or the diftillation of
" their more volatile parts, by which oily fub-
" ftances become bituminous, and bituminous
" fubftances become coally *." This, then, is

* Theory of the Earth, vol. i. p. 70.

the fuppofition which Dr. Hutton choofes to make, and which, indeed, appears to be a neceffary one for his theory, that coal is formed by heat applied to vegetable matter under different degrees of preffure, the difference in this refpect producing a difference in the refult, but that in all cafes the preffure muft be fuch as to allow of fome feparation of volatile principles.

It might be granted, that this abfence or diminution of preffure might occur in certain fituations; but how does it happen that this indifpenfable condition fhould invariably be found in the fubterranean regions, when heat is to be applied to vegetable matter. If the ufual preffure were prefent, that matter, according to Dr. Hutton, would remain unchanged with refpect to compofition by any heat ; but, in nature, we meet with no collections or ftrata of vegetable matter in this peculiar ftate, in which they have fuffered heat without being changed, yet, furely, fince heat is applied to all the other ftrata under immenfe preffure, it ought alfo to have been occafionally applied under a fimilar preffure to the matter of coal. The key to this apparent myftery is, however, eafily found. The common ftrata are fuppofed to have been heated under an immenfe preffure, which is not in any cafe fuppofed to have been abfent, *becaufe* the prefence of that preffure is neceffary, in many

cafes, to obviate certain objections to the fuppofition of their igneous origin, and, in all, an
uniformity is, if poffible, to be preferved. The
matter of coal, again, is fuppofed to have had
heat applied to it *always* under a diminifhed
preffure, *becaufe* that circumftance is neceffary,
in the Huttonian theory, to account for its formation. The bare ftatement of thefe accommodating coincidences is fufficient to prove that
they are merely fictitious, and that thefe are
fuppofitions arbitrarily made, becaufe they are
neceffary in the theory.

Another difficulty may be ftated againft this
hypothefis of the formation of coal. Pyrites,
or fulphuret of iron, is fuppofed to be formed
by heat, and this heat is fuppofed to have been
applied under fuch a preffure, as has prevented
the fulphur, which is a very volatile fubftance,
from having been driven off from the iron.
Since pyrites occur, therefore, fo abundantly in
coal, a dilemma is prefented to the Huttonian
geologift. If he fuppofe the vegetable matter
of coal to have been fufed under an immenfe
preffure, capable of refifting the efcape of its volatile parts, he contradicts the author of the
theory, who tells us, that under fuch a preffure,
that matter could not have been formed into
coal, a propofition which, according to his hypothefis, appears to be perfectly juft. If he fup-

pose the heat to have been applied under a diminished preffure, fo as to allow of the decompofition of the vegetable matter, and the expulfion of
its volatile principles, he will have the tafk of explaining how, in fuch a fituation, pyrites could
be formed and cryftallized. Nay, the difficulty
is ftill greater, for this fubftance is found in that
fpecies of coal which is little inflammable, and
which, according to the Huttonian theory, is
fuppofed to have had heat applied to it under
fo little preffure, that all its bituminous matter
had been expelled. Mr. Kirwan relates, that
the Kilkenny coal, which is of all others the
moft completely deftitute of bituminous matter,
contains pyrites* ; and Dr. Hutton himfelf men-
tions a fpecimen being in his poffeffion of plumbago, (which he confiders as the laft of the
feries, or as coal completely deprived of bitumen) ftudded with pyrites †. The explanation
of thefe appearances, according to the Huttonian fyftem, involves a direct contradiction in
terms. To account for the formation of this
fpecies of coal, it is faid to have been fufed with
an entire abfence of preffure, fo that all its bituminous matter has efcaped. Pyrites, again, is a
fubftance faid to be formed by fufion, but under a ftrong preffure, by which its fulphur, a

* Geological Effays, p. 473.   † Theory of the Earth, vol. i. p. 616.

substance at least as volatile as bitumen, is kept
in combination with the iron. It would, therefore, according to this theory, be impossible that
coal of this kind and pyrites should exist together, the circumstance supposed necessary for
the formation of the one, being that which must
inevitably have destroyed the other.

A similar argument, equally forcible, may be
deduced from the connection of limestone with
coal. No arrangement is more common in geology than that of limestone alternating with
coal, or covering it. The formation of limestone, in the Huttonian system, is ascribed to
fusion, under vast compression, by which the
carbonic acid has been retained in combination
with the lime; and the advantage which the
admission of this modifying circumstance gives
to the Huttonian geologist in his reasonings, is
very amply, perhaps ostentatiously displayed in
the defence of the system. It is always admitted, however, that but for this circumstance
of compression, the consolidation of limestone
by fusion could not be accounted for. The
formation of coal, on the other hand, it is contended, cannot take place but when pressure is
diminished or withdrawn. When limestone
covers coal, it can have been consolidated only
by the heat operating through the coal. But
if pressure were present on the limestone above,

how could it have been abfent from the coal
beneath? To fuppofe this is a palpable abfurdity ; and therefore, were the Huttonian theory true, either the coal fhould not have been
formed, or the limeftone fhould have been converted into lime.

Among the SALINE SUBSTANCES, Rock Salt is
adduced as affording an argument in favour of
the Huttonian theory. This fubftance is found
generally in ftrata, and it is perfectly compact
and indurated. This ftate, it is faid, could not
be produced by cryftallization from water ; a
mere affemblage of loofe cryftals, without folidity or cohefion, only could be formed ; and
to convert thefe into a firm and folid rock,
would require the application of fuch heat as
was able to reduce it into fufion. " The con-
" folidation of rock falt, therefore, cannot be
" explained but on the hypothefis of fubterra-
" neous heat *."

The Neptunift will find no great difficulty in
obviating this argument. If the cryftallization
has, from alteration in the circumftances of the
folution, been hafty, it is conceivable that, inftead of an affemblage of fmall regular cryftals,
large and compact cryftalline maffes might be
formed ; or even if the falt firft depofited had

* Illuftrations, &c.

not been perfectly compact, its confolidation
might be completed by the percolation of water
holding falt in folution through it, and depofiting that falt in its pores.

It feems to be altogether inconceivable, how
the formation of the immenfe ftrata of fea falt
found in nature, can be accounted for on the
Huttonian hypothefis : All its ufual principles
muft in this cafe be relinquifhed as utterly unte-
nable. It cannot be fuppofed that this falt is
derived, like the materials of all the other ftrata,
from the difintegration of an ancient world ;
for allowing that ftrata of falt exifted in that
world, and that thefe had fhared in the general
difintegration, the falt muft neceffarily have
been diffolved by the water of the ocean, to
which, in common with the other materials, it
was carried : And how, therefore, could it be
depofited, fo as to be fubjected to the action of
fubterraneous heat? This is in fact acknowledged to be a cafe which cannot be explained in
conformity to the general theory, and therefore
a different hypothefis is propofed. " If the ope-
" ration of fubterraneous heat be admitted," fays
Profeffor Playfair, " it appears poffible, that the
" local application of fuch heat may have driven
" the water in vapour from one place to another;
" and by fuch action often repeated in the fame
" fpot, may have produced thefe great accumula-

" tions of faline matter that are actually found in
" the bowels of the earth *."

This hypothefis refts on a feries of gratuitous
affumptions, fo extravagant, that, though one
were admitted, the combination of the whole,
and their adaptation to each other, may be regarded as impoffible. By what caufe, it may
be afked, was the central heat directed in its
operation to this particular fpot? Is it conceivable that any heat which it is poffible could
have thus been locally applied, would be fufficient to convert the whole water of the ocean
incumbent upon it into vapour, fo as to occafion
the precipitation of its faline matter? By what
caufe had this heat, after producing fuch an effect, ceafed to operate, fo as to allow the fea
again to cover the place from which it had been
driven? And how fhould this heat have again
been brought to act precifely on the fame fpot
with the fame force, fo as to convert the fea into vapour, and occafion a new confolidation of
its falt, and that repeated even for a number of
times? Any of thefe fuppofitions is fo romantic
as to be fufficient to invalidate any hypothefis
in which it is received; but the combination
fo far exceeds the bounds of probability, that
perhaps this might be felected as not inferior in

* Illuftrations, &c. p. 37.

extravagance to any of thofe fuppofitions which have been made in a fcience celebrated as geology has been, for the wanderings of its cultivators beyond the regions of fober reafon.

It is, perhaps, unneceffary to add thofe facts which ferve to refute fuch an hypothefis. It may be obferved, however, that were it true, rock falt fhould not contain the water of cryftallization it does; it owes its very formation to a heat capable of driving off in vapour the immenfe quantity of water in which it was diffolved, and of courfe this heat fhould have driven off the water it might be difpofed to retain. If it had been formed in this manner, the remains of marine animals fhould be abundant in it, which they are not; and laftly, the faline matter depofited fhould have been precifely that which fea water holds diffolved, muriate of foda with muriate of magnefia, and fulphate of lime. But we find rock falt fo little contaminated with thefe other falts, as to be purer even than the fea falt obtained by artificial evaporation. This laft fact at once demonftrates the falfity of the hypothefis; for it is too plain to require any illuftration, that if the fea in any part had been converted into vapour, the matter depofited muft have been a mixture of all the falts it held diffolved.

It may be difficult even on the Neptunian theory, to give a fatisfactory explanation of the

origin of rock falt; but there are no appearances
in this foffil inconfiftent with the fuppofition of
its aqueous origin. The Neptunifts have fuppofed, that it may have been formed by collections of the original fea water in hollows, among
the ftrata, having fuffered evaporation during
the temporary retreat of the ocean, and that
thefe having been fucceffively filled, have fur-
nifhed the ftrata of rock falt. But it is a ftrong
objection to this, and all other theories which
derive it from the evaporation of fea water, that
it does not contain the faline fubftances which
are diffolved in that water, or at leaft does not
contain them in the due proportion, and that it
contains fewer remains of marine animals than
we fhould expect, had it had fuch an origin.
Perhaps we may fuppofe, that the faline fubftances, in common with others, had exifted in
the original fluid in which all the materials of
the ftrata were diffolved,—that part of thefe being locally accumulated, in the fame manner as
the materials of every other ftratum have more
or lefs been, from circumftances which we cannot determine, had been confufedly cryftallized;
and that any remaining portion had been retained in folution by the water, thefe falts of courfe
remaining in it, in encreafed proportion which
were leaft difpofed to cryftallize. Such is the
muriate of magnefia; and it is not impoffible

but that even much of the muriate of foda now found in fea water, may have been rediffolved from ftrata formed. Or we may modify this explanation, by the equally probable fuppofi-tion, that at firft the foda and the muriatic acid of the falt had not been in combination, but that, in the courfe of the various alterations of attractions from the precipitations of the ftrata, they had been brought together, had united, and if locally accumulated, cryftallized. Such a fuppofition receives confirmation from the fact, that in many of the ftrata, in trap for example, according to the excellent experiments of Dr. Kennedy, both foda and muriatic acid exift; and fea falt itfelf is found fometimes among pri-mary ftrata. This general hypothefis, modified in either of thefe ways, involves no improbable fuppofitions, and is perhaps adequate to the ex-planation of the production of this foffil.

Another faline fubftance, a variety of carbo-nate of foda found in Africa, is ftated by Dr. Hutton as affording a proof of confolidation from fufion. This fubftance is fuppofed to have been part of the contents of a vein, as it has a ftony cruft adhering to one fide of it; it has a fparry ftructure, and contains only about one-fixth of the quantity of water of cryftallization contained in the ufual cryftallized ftate of this falt. It is this laft circumftance which is confidered as affording a proof that this falt has not originated from water.

We are not told in what manner this falt might be fuppofed to be formed according to the Huttonian fyftem, and it feems impoffible to give, according to the principles of that fyftem, any explanation of its origin. Every fubftance compofing any folid part of the fur-face of the globe, is fuppofed to be derived from the wafte of a former world, and to have been depofited from the fea. But fuppofe carbonate of foda to have exifted in the ftrata of the for-mer world, when diffolved in the waters, and brought to the fea, it muft have remained dif-folved, and have been diffufed through it in fuch a manner as, from the fmall quantity of it ap-parently exifting in nature, not to be capable of being difcovered; and no caufe can be ima-gined, in conformity with that fyftem, by which it could be precipitated. Even the theory which is employed for accounting for the produ&ion of fea falt, extravagant as it is, cannot be applied in the prefent cafe, for, if the fea water, holding the carbonate of foda in folution, had been ex-pofed to a local heat capable of converting it into vapour, and of thus precipitating what was diffolved in it, it is obvious that pure carbonate of foda could not have been depofited, but muft have been mixed with muriate of foda, and

Q

every other fubftance which the fea water had
held in folution ; or rather, if it ever had been
brought to the fea, it muft have immediately been
decompofed by the muriate of magnefia prefent,
and could never have exifted in the waters of
the ocean. The Huttonian theory is therefore
actually incapable of affording a fuppofition, by
which carbonate of foda, as a foffil fubftance,
could be formed. Mr. Kirwan adds, as a proof
that it has not been fufed, its containing fome
grains of fand, which would neceffarily, in fuch
a cafe, have been vitrified. And were it fuppofed to have been in fufion, it would require
to be explained how the preffure was fo nicely
adjufted as to admit of the greater part of its
water of cryftallization being driven off, while
none of its carbonic acid, even though it be
fuperfaturated with it, and of courfe retains the
excefs by a very weak affinity, had been expelled.

The production of this falt, if it really be part
of a vein, may be explained on the Neptunian
fyftem by a fimilar hypothefis to that propofed ;
to account for the formation of rock falt, and
from its hafty cryftallization, or its fuper-fatura-
tion with carbonic acid, (which, from its analyfis, is found to be the cafe,) the diminifhed
quantity of water of cryftallization in it may be
accounted for. If the fact, however, be true,
which Bergman ftates, that the carbonate of fo-

da found in the earth, both in India and Africa, is free from common falt at the furface, but becomes contaminated by it as it defcends deeper, it is probable that this falt may have been formed at the furface from the decompofition of muriate of foda by fome unknown power, and from this mode of formation its containing lefs water of cryftallization might arife.

In concluding this argument with regard to the faline fubftances, the formation of one of them which is not noticed in the Huttonian geology, gypfum or fulphat of lime may be ftated. This falt is foluble in 500 times its weight of water; of courfe, had it exifted in the ancient world, and been brought by difintegration to the ocean, it muft have remained diffolved in the water till that fluid was faturated with it; and there probably does not exift in nature a quantity fufficient for that purpofe. It would then remain to be explained how it could be collected in particular places, and be precipitated. The theory, applied to the formation of fea falt, will here be of no avail; for fhould we fuppofe it prefent in a large proportion in fea water in a particular place, the evaporation of that water could not account for its confolidation, fince it muft have been accompanied with fea falt, a fubftance of which, according to the experiments of Sauffure, gypfum frequently does not contain an atom. This, therefore, is another foſſil, for the formation of which the Huttonian theory cannot account. Its origin, on Neptunian principles, may probably be explained in the ſame manner as that of rock ſalt.

The ſtate in which the METALS are found in nature, either pure or combined with other ſubſtances, appears to afford a ſtrong argument in favour of the Huttonian hypotheſis; for the fluidity from which they have conſolidated may have been produced by fuſion, but we ſcarcely can point out by what ſolvent it could have been effected. Gold, ſilver, copper, and ſome others, are frequently found native or uncombined. " Of all ſuch ſpecimens it may be ſafe " ly affirmed, that if they have ever been fluid, " or even ſoft, they muſt have been ſo by the " action of heat; for to ſuppoſe that a metal has " been precipitated pure and uncombined from " any menſtruum is to treſpaſs againſt all analo " gy, and to maintain a phyſical impoſſibility*."

The aſſertion in this paragraph is much ſtronger than what the facts can eſtabliſh; for although it may be difficult to point out the mode in which metals have actually been precipitated, yet their precipitation from any menſtruum, ſo far from being a phyſical impoſſibility, is what happens

* Illuſtrations, &c. p. 59.

every day, and can be effected at pleasure.
Metals in a state of combination with acids,
and in solution in water, are thrown down in
their metallic state by each other, by hydrogen, sulphurated hydrogen, and various
other inflammable bodies. The Neptunian will
readily acknowledge, that it is extremely difficult to point out, even by hypothesis, by
what particular agency the metals found in nature had been dissolved and precipitated, but
at the same time he has ample demonstration,
that it is in this way, and not by fusion, that
they have been formed.

This proof is obtained from the crystallized
state in which they are frequently found. " Spe-
" cimens of quartz, containing gold and silver
" shooting through them, with the most beauti-
" ful and varied ramifications, are every where
" to be met with in the cabinets of the curious,
" and contain in their structure the clearest proof
" that the metal and the quartz have been both
" soft, and have crystallized together * "

It may be pronounced a physical impossibility, that from simple fusion quartz and gold, or
quartz and silver, could crystallize together so
as to exhibit these appearances. These metals
are fused at a heat, which, compared even with

* Illustrations, &c. p. 59.

Q 3

the degrees of heat we have it in our power to produce, may be termed very moderate, while quartz we are unable to fufe. If, therefore, both fubftances were in fufion, on a reduction of temperature, the quartz muft have become folid long before the metal; and it is abfolutely impoffible that the metal could have fhot through the quartz. Or, to place this in a point of view more precife, gold melts at a temperature equal to 32 of Wedgewood's pyrometer, and at all temperatures above this it muft remain fluid : Quartz does not melt at the higheft heat that has been accurately meafured; but, according to the experiments of Sauffure, it is not lefs than 4043 of Wedgewood's fcale, and confequently at every temperature below this, muft continue folid — Grant, therefore, to the Huttonian, that both quartz and gold were in fufion, it is evident that on a reduction of temperature to 4000, the quartz would become folid, or cryftallize ; but it is equally certain, that at this temperature, and for more than 3900 lower, the gold muft remain fluid : the fuppofition, therefore, that the gold could become folid, and fhoot through the fluid quartz, involves a direct contradiction in terms, or fuppofes a phyfical impoffibility ; and of confequence, the various appearances in thefe fpecimens which prove that the metal and quartz had cryftallized together, or that the former had

cryftallized firft, prove, as much as any phenomena can, that thefe cryftallizations could not
be from fimple fufion. It will not furely be prefumption to affirm, that if this be not admitted
as an undeniable conclufion, as affording even a
demonftration as certain as any can be, all reafoning on the fubject muft be given up, for it is
impoffible to conceive a propofition more evident, or the reverfe of which involves a more
palpable contradiction in terms. The ftatement
of thefe facts, as favourable to the Huttonian
fyftem, affords a ftriking example how far the
mind may be mifled by a favourite hypothefis,
the very appearances which prove its falfity being adduced as proofs of its truth.

One mode might perhaps occur in which it
might be attempted to remove this difficulty.
It may be conceived, that the quartz had firft
become folid, and that merely the fiffures of
it had been filled with cryftallized gold. But
fuch a fuppofition has been very clearly refuted
by Mr. Playfair himfelf. " Between the channels
" in which the metal pervades the quartz, and
" the ordinary cracks or fiffures in ftones, there is
" no refemblance whatever ; a fyftem of hollow
" tubes winding through a ftone (as the tubes
" in queftion muft have been before they were
" filled by the metal), is itfelf far more incon-
" ceivable than the thing which it is intended

" to explain ; and laſtly, if the ſtone was per-
" forated by ſuch tubes, it would ſtill be infi-
" nite to one that they did not all exactly join,
" or inoſculate with one another *."

The concluſion which may be drawn from
this argument is ſomewhat ſingular : The ſtronger the objection is ſtated, and it cannot be more
ſtrongly urged than it has been by Profeſſor
Playfair, the more favourable on the whole is
it to the Neptunian ſyſtem. Let it be placed
in the cleareſt light, ſuppoſe it even to be ſo
ſtrongly urged, that the Neptuniſt is unable to
give any probable conjecture as to the mode in
which the metals have. been formed by ſolution ; what is the fair concluſion ? It is, that difficulties of this kind are inſeparable from the
ſubject, or rather from our imperfect knowledge,
and are therefore comparatively of little importance, if they do not involve inconſiſtencies with
the principles of the theory, or with eſtabliſhed
facts. From the appearances of the metals, from
the diſſemination of cryſtallized gold in quartz
alone, we have a clear and unexceptionable demonſtration that they do not owe their origin
to fuſion. No other mean can be pointed out,
or even imagined, but ſolution ; it has already
been pointed out, how from ſolution ſimultane-

* Illuſtrations, &c. p. 245.

ous confolidation might take place; and it is fufficiently obvious, that though quartz is lefs fufible than gold, and muft therefore confolidate
before it from *fufion*, it might be more foluble
in the menftruum in which both were diffolved,
and might therefore remain fluid while the gold
cryftallized from *folution*. But were it even
impoffible to conjecture how they fhould have
confolidated from folution, fo as to produce the
appearances obferved, the conclufion muft ftill
be drawn that they had been formed in this
mode ; for on the one hypothefis we fhould
have only a deficiency of explanation, in the
other, a direct and pofitive contradiction to an
eftablifhed truth.

This proof of the aqueous origin of metals is
therefore capable of being carried a great length,
fince there are certainly no foffils which at firft
view would appear lefs likely to have been
formed by water. It alfo of itfelf eftablifhes a
fimilar origin to almoft every other mineral, for
the metals are fo intimately connected with fo
many of them, with quartz, carbonate, and fluate
of lime, fulphate of barytes, and many others,
that whatever has been the origin of the one,
muft have been the origin of the other.

There are fome other facts refpecting the native metals, cited as proofs of the Huttonian
theory, particularly the large fpecimens of iron

found native in Siberia and Peru. The moſt remarkable circumſtance with regard to theſe maſſes is their largeneſs. That found in Siberia weighs 15 tons, and is ſoft and malleable. The American ſpecimen is alſo very large, and there is a peculiar appearance connected with it,—the impreſſions of the feet of men and birds on its ſurface. Theſe maſſes are concluded to have been formed by fuſion, and to have been part of the contents of a vein waſted away, from which, the iron being the more durable ſub-ſtance has been left on the ſurface of the ground.

In conſidering the ſingular circumſtances at-tending theſe maſſes, what would one conclude who was guided by the common rules of rea-ſoning. He would infer merely that their ori-gin was at preſent inexplicable. The defender of the Huttonian theory purſues a different mode; becauſe it is inconceivable, as he ima-gines, that they could have been formed by precipitation from ſolution, he therefore con-cludes, that they had been formed by fire, and brings them forward as proofs of his ſyſtem. With regard to the very peculiar circumſtance attending one of theſe maſſes we receive no information; we are not told how birds and men found their way to the central regions, and left the impreſſions of their feet on the fuſed iron, but the obfervation is merely made, that fuch circumftances " are not to be accounted for on " any hypothefis, and certainly require more " careful inveftigation *." The conclufion would have been more accurate that they demonftrate that thefe maffes could not have been formed by fufion and injection. The fact might be ftated as a fingular one in a fyftem of natural hiftory, but in reafoning on a theory of the earth, it ought while involved in fuch obfcurity to have found no place, far lefs fhould it ever have been brought forward as a proof of the fu-fion of minerals.

The native combination of metals with ful-phur, affords another argument to the Hutto-nian geologift. It is obferved, that neither the metal nor the fulphur is foluble in water; and that the metallic fulphuret, even when formed, is decompofed by water, while, on the other hand, fulphur, and the greater number of the metals, can be fufed and combined by heat. It is hence concluded, that pyrites, and other na-tive combinations of fulphur with metals, muft have been formed by fufion, and this conclufion has been employed as an argument to prove the igneous origin of a number of foffils in which thefe fulphurets are found.

* Illuftrations, &c. p. 240.

The argument, *a priori*, that pyrites cannot be formed in the humid way, becaufe neither the fulphur nor the metal is foluble in water, is founded on an erroneous conclufion; for, granting the fact, each might have been combined with other fubftances which would render them foluble, and, in this ftate of folution, they might, from the chemical affinity fubfifting between them, leave the fubftances with which they were united, and combine together. This actually happens in the example of fulphurated hydrogen, with a number of the metallic falts or oxyds. If this compound of hydrogen and fulphur, be introduced into a folution of any of the falts of lead, the hydrogen combines with the oxygen of the metallic oxyds, and the lead combines with the fulphur, forming a compound which, according to the obfervation of Vauquelin, has all the propeities of galena. It is not even neceffary that the metal fhould be combined with an acid, for if humidity be prefent it will be oxydated, and upon this oxyd the fulphurated hydrogen is capable of acting and producing a metallic compound, or the fame fubftance is even capable of acting on the pure metals.

This argument muft be admitted as perfectly conclufive, when it is proved, that pyrites, and other compounds of the metals with fulphur, are formed by nature in the humid way, and of this the proofs are abundant.

Thus pyrites is found in fituations which clearly indicate its aqueous formation. It is affociated with calcareous cryftals lining the internal cavity of fhells, which could not have been in fufion, fince the heat neceffary to fufe the pyrites, or the calcareous cryftals, muft have deftroyed the texture of the fhell,—it is often to be traced in the impreffions of organic fubftances, particularly of animal remains in coal and other foffils; it exifts in bituminated wood, which, it muft be granted, could not have been fufed; and it has been found forming on the furface of wood, in mines, and in other fituations, which eftablifh the fame conclufion, of which feveral examples are ftated by Mr. Kirwan *.

The obfervations of Mr. Wifeman on the effects of the waters of the Mere of Difs on metallic fubftances; and the farther experiments and obfervations of Mr. Hatchet, on the fame fubject, are alfo particularly valuable in proving the humid formation of thefe compounds. Mr. Wifeman obferved, that flints, and other ftones, immerfed for fome time in this ftagnant water,

* Geological Effays, p 400

were incrufted with a metallic ftain. This, by
analyfis, was found to be fulphuret of iron; and
when copper was kept in this water, it was encrufted by a fubftance which was found to be
compofed of 70 parts of copper, 16. 6 of fulphur, and 13. 3 of iron. This encruftation was
even found in a cryftallized ftate. Thefe experiments on this fubftance, and on the encruftation
of martial pyrites on the flints, were confirmed
by Mr. Hatchet. With refpect to the latter,
he obferves, there could be no hefitation; and
the former he confidered as " in every property
" fimilar to that rare fpecies of copper ore, call-
" ed by the Germans *Kupfer fchwärtze*, (cu-
" prum ochraceum nigrum) and abfolutely the
" fame." At the defire of Mr. Hatchet, filver
was immerfed in this water, and it was found to
be encrufted with a fubftance " fimilar in every
" refpect to the fulphurated or vitreous ore of
" filver, called by the Germans, Glafertz."
The fame diftinguifhed chemift adds, that effects
fimilar to thefe, on a larger fcale, there is reafon
to believe, " have been, and are now daily pro-
" duced in many places. The pyrites in coal
" mines have probably, in great meafure, thus
" originated. The pyritical wood may thus
" have been produced; and by the fubfequent
" lofs of fulphur, and orydation of the iron, this

" pyritical wood appears to have formed the
" wood like iron ore, which is found in many
" parts *." Of the formation of metallic fulphurets in the humid way, there can therefore
be no doubt, and inftead of having any difficulty
to encounter in explaining the origin of thefe
fubftances, the Neptunift may juftly bring them
forward, as affording a proof of his theory, and
a proof of confiderable importance, from the
extenfive connections they have with other
foffils.

If it were neceffary to fay more on this fubject, which perhaps it is not, it might be remarked, that from the manner in which thefe compounds are affociated with certain foffils, we
have the fame demonftration as in the example
of the pure metals, that they cannot have been
formed by fufion. Not only the fulphuret of
iron, but thofe of antimony, mercury, and filver,
are frequently found cryftallized or diffeminated
through quartz and other foffils. Now thefe
fulphurets are very eafily fufed, while thefe foffils are comparatively infufible ; it is therefore
impoffible that the former eould have cryftallized within the latter, or been diffeminated through
them, fince, to admit of its regular cryftallization, or even of its diffemination in another

body, that body muft have afforded little or no refiftance, or been entirely fluid, or extremely foft ; but quartz could not be fluid from fufion, and at the fame time fulphuret of antimony or mercury have been folid.

The ftruĉture and appearances of GRANITE as a foffil, have been brought forward by Dr. Hutton as favourable to his hypothefis. There can be no doubt that this rock has at one time been fluid. Its conftituent parts, particularly the felfpar, and fometimes the quartz, are chryftallized, and it not unfrequently contains other cryftallized foffils. This fluidity from which it has been confolidated, Dr. Hutton conceives is proved to have been that of fufion from the parts of the granite impreffing each other. The fpecies of granite termed Graphic, is ftated, particularly as eftablifhing this deduĉtion. In it the felfpar is cryftallized in its ufual rhomboidal form, and thefe cryftals imprefs the quartz, put it afide as it were, and give it its particular fituation a-long the fides of the rhomboidal felfpar. " Hence " this granite is not a congeries of parts, which, " after being feparately formed, were fomehow " brought together and agglutinated, but it is " certain that the quartz at leaft was fluid when " it was moulded on the felfpar." And " this " fluidity was not the effeĉt of folution in a " menftruum, for in that cafe one kind of cry-

" ſtal ought not to impreſs another, but each
" of them ſhould have its own peculiar ſhape*."

It has already been ſhown, how ſimultaneous
conſolidation may take place from ſolution ;
and the ſtructure of granite can, from this circumſtance, furniſh no argument againſt its aqueous origin. But it affords the cleareſt demonſtration, that it has not been formed by fuſion.
Felſpar is a ſubſtance incomparably more fuſible
than quartz, the one varying from 120 to 150,
the other being 4043 of Wedgewoods ſcale.
It is a propoſition, therefore, ſelf evident and
undeniable, that in the ſame maſs quartz could
not be fluid when felſpar was ſolid ; and therefore,
ſince in this graphic granite the quartz is moulded on the cryſtallized felſpar, (and, in the greater number of granites, the felſpar is cryſtallized
while the quartz is not) the fluidity whence
both have been conſolidated cannot have been
fuſion from heat.

The force of this argument, from the cryſtallization of felſpar in granite, may be eſtimated
by a very ſimple conſideration. If the appearance of this foſſil had been the reverſe of what
it is, if the quartz had cryſtallized and ſhot
through the felſpar, would it not have been
brought forward as a proof, or at leaſt as a ſtrong

---

* Illuſtrations, &c. p. 86.

R

prefumption, that both had confolidated from
fufion? Yet the fact, as it is, is ftill ftronger in
proving, that fuch could not have been their
origin, fince it is an evident impoffibility, that,
in two fubftances fufed together, the moft fufible
fhould concrete before the one that was leaft
fufible, or required the greateft heat to keep it
fluid. Had the former even been congealed by
fome caufe, of which we can form no conception,
the high temperature keeping the other, fluid,
muft have again immediately fufed it.

Under this head may be noticed feveral facts,
affording an argument of a fimilar kind ;—thofe
in which foffils are regularly cryftallized, in
others which are much lefs fufible. Thus befides the cryftallization of felfpar in granite, regular cryftals of it are not unfrequently found
embedded in quartz, a proof which cannot be
eluded, that the felfpar has become folid, while
the quartz remained fluid, contrary to what
muft have happened from the known fufibilities
of thefe fubftances, if they had confolidated from
fufion. Shorl is a fubftance of comparatively
eafy fufibility, yet it is often cryftallized in
quartz, the fibres of the fhorl being finer even
than the human hair, fhooting through a large
mafs of quartz in every direction, and with various wavings and incurvations, fo as to prove
decifively that the quartz had been completely

liquid when the fhorl cryftallized. Afbeftos is
another fubftance found fhooting in the moft
delicate fibres through quartz, though it melts
at 378 of Wedgewood's fcale. And, without
enumerating more examples, micaceous fhiftus
is a rock of very difficult fufion, yet it is the
common matrix in which garnets are enveloped, and though thefe are much more fufible,
they are cryftallized in its fubftance with the
greateft regularity. It muft be fuperfluous to
repeat the argument from fuch facts. If they
do not prove that thefe foffils have not been
formed by fufion, no conclufion can be eftablifhed in geology, and we may relinquifh every attempt to theorife.

Whin or trap is a rock into which granite
infenfibly graduates, fo that what is proved with
refpect to the one, may be confidered nearly as
proved with regard to the other. There are
fome particular facts, however, with regard to
whin, which are fuppofed by the defenders of
the Huttonian theory to prove its igneous origin.

Whin, it is faid, refembles lava in its appearance fo much, that fome varieties of it have been
miftaken for volcanic products. This refemblance " leads to fufpect, that the two ftones
" have the fame origin, and that as lava is cer-
" tainly a production of fire, fo probably is

R 2

" whinftone *." Any diverfity exifting between them confifts principally in whinftone containing calcareous fpar, which lavas do not ; and
this diverfity is likewife explained in the Huttonian fyftem ; as this fubftance might be formed
in whin, from the fufion of it having been under
an immenfe preffure, while lava, being in fufion
expofed to the air, it muft have been decompofed. " Thus," it is added, " whinftone is to be
" accounted a fubterraneous or *unerupted* lava ;
" and our theory has the advantage of explain-
" ing both the affinity and the difference be-
" tween thefe ftony bodies, without the intro-
" duction of any new hypothefis. In the Nep-
" tunian fyftem, the affinity of whinftone and
" lava is a paradox which admits of no folu-
" tion †."

To this argument it may be replied, that the
mere refemblance in appearance between different foffils is a very weak proof of a fimilar origin, fince in many cafes clofe refemblances are
to be traced between foffils altogether different
in their nature. But were it a juft conclufion in
general, in the prefent cafe it is not fo, becaufe
the refemblance can be otherwife very eafily explained. It is extremely probable that lava

* Illuftrations, &c. p. 68.
† Illuftrations, &c. p. 69.

is formed from rocks of the nature of whin, fufed by the volcanic fire. It is apparent, from the defcriptions of Spallanzani, and other mineralogifts, that rocks of this fpecies are the bafis of volcanic countries: zeolite, leucite, and other foffils, ufually found in rocks of this order, are likewife contained in lavas, altered, but, according to the opinion of the beft mineralogifts, not formed by the volcanic fire. And, laftly, the excellent analyfes of Dr. Kennedy prove the near refemblance in compofition of trap and lava. They fhow, as he himfelf obferves, " that whins, and a certain clafs of lavas taken " from remote quarters of the globe, confift of " the fame component elements united in each, " nearly in the fame proportion. The only cir- " cumftance in which they materially differ, is " the lofs of fome volatile matter in the fire, " which is peculiar to the whins alone." The conclufion, therefore, evidently amounts even to more than a high probability that lava has been formed from the fufion of trap; and if this be true, the refemblance between them is no proof whatever of their having had a fimilar origin. Whatever may have been the origin of the trap,—although it be of aqueous formation, it is perfectly conceivable, that if fufed, as none of its principles are loft, it might form a fub-

ftance very fimilar in its properties to the trap
in its original ftate. It is equally evident, that
the difference between trap and lava, that of
the latter containing no carbonate of lime, is on
this fuppofition fully explained, fince, if the trap
were fufed in an open volcanic fire, the carbonic acid would be expelled from the lime. The
affinity, therefore, between lava and whinftone,
in fome points, and their difference in others,
are fully accounted for, and cannot be regarded
as forming, " in the Neptunian fyftem, a para-
" dox which admits of no folution."

But there is a difference between thefe fubftances not noticed by the Huttonian geologift,
and for which he will find it difficult to account.
We are told that trap differs in lava, in nothing
but in the circumftances of their formation, the
one having been melted matter erupted at the
furface while fluid, the other having been thrown
up among folid ftrata, and confolidated under an
immenfe preffure. Hence is explained the prefence of carbonate of lime in the one, while it
is not found in the other ; and it follows, from
the opinion itfelf, that lava and trap fhould differ from each other in nothing but in fuch properties or appearances as are capable of being
produced by the caufe thus fpecified. Will the
Huttonian geologift then inform us why agates,

and maſſes or veins of quartz, or even regular
cryſtals of it, which are abundant in baſalt, are
not to be found in lava? The abſence of compreſſion could not prevent their formation, or
render them more ready to be deſtroyed if they
were formed ; and it is evident that lava and
trap ought, according to the Huttonian theory,
to differ in nothing but what this abſence of
compreſſion in the one caſe could occaſion.

The columnar ſtructure which the cloſe-grain-
ed whin ſometimes aſſumes, forming baſalt, has
been ſuppoſed a proof of its igneous origin, becauſe the ſame ſtructure, it is ſaid, is ſometimes
aſſumed by the lava actually erupted from vol-
canos. It is to be obſerved, that in the greater
number of lavas, cooled under every variety of
ſituation, either ſlowly by expoſure to the air,
or rapidly by having flowed into the ſea, this
columnar appearance is not to be obſerved ; and
many of the inſtances which have been given
are extremely doubtful, from rocks not volcanic
having been ſo often confounded with lavas. If
it be admitted that real lavas do ſometimes aſſume a columnar form, (and ſuch inſtances, if
they exiſt, are extremely rare,) ſtill theſe facts
clearly ſhow that this has not been an effect ariſing from their ſpecies of fluidity, or their mode
of conſolidation, as in by far the greater number of caſes in which theſe cauſes muſt have

R 4

equally operated, it is abfent. Did it arife, indeed, from either of them, it ought to be obferved in almoft all the varieties of whin, and
might even be expected in the other unftratified
rocks. It may therefore be afcribed with more
probability to another caufe ;—the peculiar nature or compofition of the matter of which lava
confifts : And if this, as there is every reafon to
believe, be the real caufe, it is evident, that fince
bafalt is perfectly fimilar in compofition to lava,
it might affume the fame form in its confolidation from an aqueous origin : that, in fhort, if
the form do not originate from the mode of confolidation, but from a property belonging to the
fubftance itfelf, and originating in its compofition, it may equally be exhibited by lava becoming folid from fufion, and bafalt becoming
folid from folution. It is not improbable that
this property may arife chiefly from the predominance of argil in thefe foffils, as it is found to
take place even in other argillaceous foffils, as
in the argillaceous iron ore, fome varieties of
marl, and even fome of argillaceous fandftone,
—fubftances, fome at leaft of which have evidently never been fufed.

" A mark of fufion, or at leaft of the opera-
' tion of heat, which whinftone poffeffes in com-
" mon with many other minerals, is its being
" penetrated by pyrites—a fubftance, as has been

already remarked, that is of all others moſt ex-
cluſively the production of fire *." The ſuperior probability of pyrites being of watery origin has been already ſhown, and of courſe this
fact becomes a ſtrong argument for the aqueous
formation of whin. A ſimilar argument is that
derived from the preſence of agates in trap, theſe
foſſils being ſuppoſed to give indications of
having been formed by fire. This ſuppoſition
has already been conſidered, and the concluſion
endeavoured to be proved, that, in common with
other foſſils, they owe their formation to the
agency of water.

The laſt argument for the igneous origin of
whin, which has not been noticed, is that deduced from the experiments of Sir James Hall.
It had been often ſtated, as an objection to the
opinion that baſalt was of igneous origin, that
it ought to have a vitreous luſtre and fracture, ſince from melting any earthy combination ſome ſpecies of glaſs is always produced,
and ſince baſalt itſelf by fuſion actually forms a
real glaſs. Sir James Hall, by a number of excellent experiments, has clearly ſhown, that
when this ſtone is brought into fuſion, if the re-
frigeration of it be very ſlow, it aſſumes the
ſtony character, or is ſcarcely diſtinguiſhable

* Illuſtrations, &c.

from the real bafalt, and that it is only when
haftily cooled that it acquires any of the properties of glafs.

By fome of the defenders of the Huttonian
theory the conclufions from thefe experiments
have been carried much farther than they warrant, or than their author has ftated. Becaufe from
the fufion of trap or bafalt a fubftance fimilar to
thefe foffils was by particular management produced, it has been concluded that this affords
a proof of fufion being the means of their firft
formation. This conclufion is an evident miftake : the fufion of the bafalt, as it contained
no volatile fubftance of importance, could not
alter its compofition, and of courfe when it again
became folid, it would ftill be endowed with its
former properties. Even if the fimple earths
of which bafalt confifts had been mixed together in the proportions which its analyfis affords—if by fufion thefe had been brought into
union, and formed a fubftance fimilar to natural bafalt, ftill this would not have proved the
igneous formation of this foffil ; for as all compounds derive their properties from their compofition, if their conftituent parts are capable
of being brought into union by the humid way,
and alfo by fufion, it may be expected that a
fimilar compound will in both cafes be formed ;
and the actual production of fuch a compound

by one of thefe modes, would not prove that it was incapable of being formed in the other. Had therefore this experiment been made, no conclufion of this kind could have followed, ftill lefs can it be inferred from merely fufing natural bafalt, and finding that by flow cooling a fubftance fimilar to it is formed. In ftrict reafoning, this experiment adds nothing pofitive to the evidence of the Huttonian fyftem ; it only removes an objection which could have been urged with juftice againft it,— and in this point of view it may be prized by the defender of that doctrine.

It is remarked, that " the experiments of ano-" ther ingenious chemift, Dr. Kennedy, have " fhown, that whinftone contains mineral alkali, " by which of courfe its fufion muft have been " affifted*." It may be added, that the prefence of this alkali would not lefs contribute to the folubility of the matter of whin in water.

Befides the proofs of the aqueous origin of whin, which have been noticed in the courfe of this argument, it may be added, that there are facts which indifputably eftablifh it. Thefe are, the exiftence of foreign bodies in it, which could not poffibly have been prefent had it been thrown up in a ftate of fufion from the

* Illuftrations, &c. p. 80.

bowels of the earth. Mr. Jamiefon has given a
very full enumeration of thefe, on authorities
which it would be prefumption to difpute. Thus,
Werner found in wacken, great trees with branches, leaves, and fruit, and obferves, that it is
fometimes found to contain deers horns. Sauffure obferved in it bones of quadrupeds; and
other mineralogifts have found fhells, vegetable
impreffions, and fragments of wood. It has
been ftated with refpect to fome of thefe, that
they are found, not in the bafalt, but in ftrata
which alternate with it; and in fome cafes this
appears to have been the cafe. But in the obfervations of Werner and Sauffure we are not
at liberty to fuppofe that they committed fo
obvious a miftake; and therefore thefe obfervations are decifive proofs of the aqueous origin
of thefe rocks.

The properties therefore of the fubftances
compofing the unftratified rocks equally with
their pofitions exclude the operation of fire,
and prove them to be of aqueous origin.

We have thus completed the examination of the
HUTTONIAN and NEPTUNIAN theories; and it cannot be difficult to form an opinion on their merits.
To the Huttonian fyftem belongs the praife of
novelty, boldnefs of conception, and extent of
views. Its author has afpired not merely to ac-

count for the prefent appearances of the earth, but to trace a fyftem in which the formation of fucceffive worlds is developed; he has fought to extend that order and arrangement, that principle of balance and reftoration obferved in all the departments of nature, to the conftitution of the globe itfelf; and he has fucceeded in drawing an outline which gratifies the imagination with the femblance of grandeur and defign.

But thefe are the only merits of the theory, and they have certainly been much over-rated by the partiality of its defenders. When full fcope is given to the imagination, when the reftraint of ftrict induction is not impofed, experience, and efpecially experience in geology, has fhown, that it is no difficult tafk to conftruct a fyftem, and to give it in appearance that unity of principle, and adaptation of parts which are the attributes of a perfect theory. This may be what the author of the Huttonian fyftem has attained; but a more juft and cautious reafoner would have ftartled at the *poftulata* the fyftem requires, and would not have thought their extravagance and improbability, their inconfiftency with phenomena, and their oppofition to eftablifhed truths, compenfated by the greatnefs or novelty of its views.

It can fcarcely be neceffary to juftify thefe obfervations by any recapitulation of the evidence of this theory. In appealing to the proof from induction, we have found the phenomena of geology entirely at variance with its principles. It has not been neceſſary to ſearch for minute deficiences, or ſelect a few concluſive arguments from a number that are doubtful or obſcure ; the whole ſeries is clear and convincing ; the poſitions and relations of the great maſſes of the globe, and the properties and appearances of individual foſſils, being altogether incompatible with the ſuppoſition of their having been formed by a central fire. Its firſt principles we have found not merely in the higheſt degree improbable, but abſurd and phyſically impoſſible. It aſſumes the exiſtence of an intenſe heat in the interior parts of the earth, without aſſigning any cauſe by which it could have been produced ; it ſuppoſes an exertion of that heat, not merely at the formation, or during the period of the exiſtence of a world, but for a time abſolutely unlimited ; and it advances theſe ſuppoſitions in expreſs contradiction to the known and eſtabliſhed laws of the power it employs. Such characters bring the Huttonian hypotheſis under the ſame claſs with thoſe geological ſyſtems which have preceded it,—ſyſtems which have been the meteors of their day, and have ſunk into obſcurity ; and which, to uſe the language of Dolomieu, will never be mentioned

in the hiſtory of the ſcience, but as pointing out
the by-paths in which thoſe may wander who
devote themſelves to the contemplation of nature.

With this ſyſtem, the Neptunian theory forms
a complete contraſt. It preſumes not to carry
its reſearches paſt the commencement of the preſent world, or to extend them beyond its termination ; it is ſatisfied with endeavouring to
trace the cauſes of the appearances which at
preſent exiſt ; and the characters of its explanations are theſe of fair and legitimate deduction.
All the phenomena of geology conſpire to prove
that water has been the great agent by which
minerals have been formed, and the ſurface of
the earth arranged. While the ſcience remains
in an imperfect ſtate, deficiencies muſt be found
in the application of this principle which induction eſtabliſhes. Such there may be in the
Neptunian theory ; and that there ſhould, is
even a preſumption of its truth. But we diſcover no inconſiſtencies with that principle, nor
contradictions to known truths. We find in it,
in reality, what we ſhould at preſent expect in a
juſt theory of the earth : not the magnificent pretenſions of an artificial ſyſtem,—not the
ſplendid illuſions created by a bold imagination,
but a ſeries of inductions more or leſs perfect,
referred to a common principle, and occaſional-

ly connected by a moderate and rational hypothefis. In a word, it may be confidered as the
commencement of a fyftem which poffeffes the
ftability of truth and which time will extend
and improve.

In this Comparative View the author has endeavoured to conduct the difcuffion with fair-
nefs and candour : he is not confcious of having
fuppreffed a fingle argument of importance, or
of having ftated any fo as to difguife its
ftrength : And he trufts, that, in fupporting the
fyftem he defends, he has not been wanting in
that refpect fo juftly due to the diftinguifhed
author of the " Illuftrations of the Huttonian
" Theory."

Printed by Mundell and Son, Edinburgh.

Printed in the United States
By Bookmasters